DATA ANALYSIS

DATA ANALYSIS

What Can Be Learned From the Past 50 Years

Peter J. Huber
Professor of Statistics, retired
Klosters, Switzerland

WILEY

A JOHN WILEY & SONS, INC., PUBLICATION

Published by John Wiley & Sons, Inc., Hoboken, New Jersey
Published simultaneously in Canada

Library of Congress Cataloging-in-Publication Data:

Huber, Peter J.
 Data analysis : what can be learned from the past 50 years / Peter J. Huber.
   p. cm. — (Wiley series in probability and statistics ; 874)
 Includes bibliographical references and index.
 ISBN 978-1-118-01064-8 (hardback)
 1. Mathematical statistics—History. 2. Mathematical statistics—Philosophy. 3. Numerical
analysis—Methodology. I. Title.
 QA276.15.H83 2011
 519.509—dc22   2010043284

Printed in the United States of America.

10 9 8 7 6 5 4 3 2 1

CONTENTS

PREFACE

"These prolegomena are not for the use of apprentices, but of future teachers, and indeed are not to help them to organize the presentation of an already existing science, but to discover the science itself for the first time." (Immanuel Kant, transl. Gary Hatfield.)

"Diese Prolegomena sind nicht zum Gebrauch vor Lehrlinge, sondern vor künftige Lehrer, und sollen auch diesen nicht etwa dienen, um den Vortrag einer schon vorhandnen Wissenschaft anzuordnen, sondern um diese Wissenschaft selbst allererst zu erfinden." (Immanuel Kant, 1783)

How do you learn data analysis?

First, how do you teach data analysis? This is a task going beyond "organizing the presentation of an already existing science". At several academic institutions, we tried to teach it in class, as part of an Applied Statistics requirement. Sometimes I was actively involved, but mostly I played the part of an interested observer. In all cases we ultimately failed. Why?

I believe the main reason was: it is easy to teach data analytic techniques, but it is difficult to teach their use in actual applied situations. We could not force the students to immerse themselves into the underlying subject matter. For the students, acquiring the necessary background information simply was too demanding an effort, in particular since not every subject matter appeals to every person. What worked, at least sort of, was the brutal approach we used in the applied part of the Ph.D. qualifying exam at Harvard. We handed the students a choice of some non-trivial data analysis problems. We gave them some general advice. They were asked to explain the purpose of the study furnishing the data, what questions they would want to answer and what questions they could answer with the data at hand. They should explore the data, decide on the appropriate techniques and check the validity of their results. But we refrained from giving problem-specific hints. They could prepare themselves for the exam by looking at problems and exemplary solutions from previous years. Occasionally, the data had some non-obvious defects – such as the lack of proper matching between carriers and controls, mentioned in Section 2.3. The hidden agenda behind giving those problems was not the surface issue of exam grades, but an attempt to force our students to think about non-trivial data analysis issues.

Clearly, the catalog of tools a data analyst can draw upon is immense – I recall John Tukey's favorite data analytic aphorism: *All things are lawful for me, but not all things are helpful* (1 Cor 6.12). Any attempt to cover in class more than a tiny fraction of the potentially useful tools would be counterproductive. But there are systematic defects and omissions – some of which had been criticized already back in 1940 by Deming. The typical curricula taught in statistics departments are too sanitized to form a basis for an education in data analysis. For example, they hardly ever draw attention to the pitfalls of Simpson's paradox. As a student, you do not learn to use general nonlinear fitting methods, such as nonlinear least squares. Methods that do not produce quantitative outcomes (like P-values), but merely pictorial ones, such as correspondence analysis or multidimensional scaling, as a rule are neglected.

As a data analyst, you often will have to improvise your own approaches. For the student, it therefore is indispensable to become fluent in a computer language suitable for data analysis, that is, a programmable command language (sometimes, but not entirely accurately, called a script language).

I worry about an increasing tendency, which nowadays seems to affect university instruction quite generally. Namely, instead of fostering creative improvisation and innovation by teaching the free use of the given means (I am paraphrasing Clausewitz, quoted at the opening of Section 2.2.1), one merely teaches the use of canned program packages, which has the opposite effect and is stifling.

But to come back to the initial question: how do you learn data analysis? In view of our negative experiences with the course work approach, I believe it has to be learned on the job, through apprenticeship and anecdotes rather than through systematic exposition. This may very well be the best way to teach it; George Box once pointedly remarked that you do not learn to swim from books and lectures on the theory of buoyancy (Box 1990). The higher aspects of the art you have to learn on the job.

But there are deeper questions concerning the fundamentals of data analysis. By 1990 I felt that I knew enough about the opportunities and technicalities of interactive data analysis and graphics, and I drifted into what you might call the philosophy of data analysis. Mere concern with tools and techniques is not sufficient, it is necessary to concern oneself about their proper use – when to use which technique – that is, with questions of strategy.

The advent of ever larger data sets with ever more complicated structures – some illustrative examples are discussed in Section 3.8 – forces one to re-think basic issues. I became upset about the tunnel vision of many statisticians who only see homogeneous data sets and their exclusive emphasis on tools suited for dealing with homogeneous, unstructured data. Then there is the question of how to provide suitable computing support, to permit the data analyst the free and creative use of the available means. I already mentioned that script languages are indispensable, but how should they be structured? Finally, I had again to come back to the question of how one can model unsanitized data, and how to deal with such models.

The central four chapters of this book are based on four workshop contributions of mine concerned with these issues: strategy, massive data sets, computing languages, and models. The four papers have stood the test of time remarkably well in this fast-living computer world. They are sprinkled with examples and case studies. In order that the chapters can be read independently of each other, I decided to let some overlaps stand.

The remaining chapters offer an idiosyncratic choice of issues, putting the emphasis on notoriously neglected topics. They are discussed with the help of more examples and case studies, mostly drawn from personal experiences collected over the past 50 years. Most of the examples were chosen to be small, so that the data could be presented on a few printed pages, but I believe their relevance extends far beyond small sets. On purpose, I have presented one large case study in unsanitized, gory detail (Section 5.7). Chapter 6 is concerned with what I consider some of the most pernicious pitfalls of data mining, namely Simpson's paradox (or the neglect of inhomogeneity), invisible missing values, and conceptual misinterpretation of regression. Chapter 7 is concerned with techniques to create order in the data, that is, with techniques needed as first steps when one is confronted with inhomogeneous data, and Chapter 8 is concerned with some mixed issues, among them with dimension reduction through nonlinear local modeling, including a brief outline of numerical optimization. I believe that the examples and case studies presented in this book cover in reasonable detail all stages of data analysis listed in Section 2.5, with the exception of the last: "Presentation of Conclusions". The issues of the latter are briefly discussed in Section 2.5.9, but they are not very conducive for presentation in form of a case study. The case studies also happen to exemplify applications of some most helpful tools about whose neglect in the education of statistics students I have complained above: the singular value decomposition, nonlinear weighted least squares, simulation of stochastic models, scatter- and curve plots.

Finally, since most of my work as a mathematical statistician has been on robustness, it is appropriate that I state my credo with regard to that topic. In data analysis, robustness has pervasive importance, but it forms part of a general diligence requirement and therefore stays mostly beneath the surface. In my opinion the crucial attribute of robust methods is stability under small perturbations of the model, see in particular the discussion at the beginning of Section 5.1. Robustness is more than a bag of procedures. It should rather be regarded as a state of mind: a statistician should keep in mind that *all* aspects of a data analytic setup (experimental design, data collection, models, procedures) must be handled in such a way that minor deviations from the assumptions cannot have large effects on the results (a robustness problem), and that major deviations can be discovered (a diagnostic problem). Only a small part of this can be captured by theorems and proofs, or by canned computer procedures.

PETER J. HUBER

Klosters
December 2010

CHAPTER 1

WHAT IS DATA ANALYSIS?

Data analysis is concerned with the analysis of data – of any kind, and by any means. If statistics is the art of collecting and interpreting data, as some have claimed, ranging from planning the collection to presenting the conclusions, then it covers all of data analysis (and some more). On the other hand, while much of data analysis is not statistical in the traditional sense of the word, it sooner or later will put to good use every conceivable statistical method, so the two terms are practically coextensive. But with Tukey (1962) I am generally preferring "data analysis" over "statistics" because the latter term is used by many in an overly narrow sense, covering only those aspects of the field that can be captured through mathematics and probability.

I had been fortunate to get an early headstart. My involvement with data analysis (together with that of my wife who then was working on her thesis in X-ray crystallography), goes back to the late 1950s, a couple of years before I thought of switching from topology into mathematical statistics. At that time we both began to program computers to assist us in the analysis of data – I got involved through my curiosity in the novel tool. In 1970, we were fortunate to participate in the arrival of

Data Analysis: What Can Be Learned From the Past 50 Years. By Peter J. Huber
Copyright © 2011 John Wiley & Sons, Inc.

non-trivial 3-d computer graphics, and at that time we even programmed a fledgling expert system for molecular model fitting. From the late 1970s onward, we got involved in the development and use of immediate languages for the purposes of data analysis.

Clearly, my thinking has been influenced by the philosophical sections of Tukey's paper on "The Future of Data Analysis" (1962). While I should emphasize that in my view data analysis is concerned with data sets of any size, I shall pay particular attention to the requirements posed by large sets – data sets large enough to require computer assistance, and possibly massive enough to create problems through sheer size – and concentrate on ideas that have the potential to extend beyond small sets. For this reason there will be little overlap with books such as Tukey's *Exploratory Data Analysis* (EDA) (1977), which was geared toward the analysis of small sets by pencil-and-paper methods.

Data analysis is rife with unreconciled contradictions, and it suffices to mention a few. Most of its tools are statistical in nature. But then, why is most data analysis done by non-statisticians? And why are most statisticians data shy and reluctant even to touch large data bases? Major data analyses must be planned carefully and well in advance. Yet, data analysis is full of surprises, and the best plans will constantly be thrown off track. Any consultant concerned with more than one application is aware that there is a common unity of data analysis, hidden behind a diversity of language, and stretching across most diverse fields of application. Yet, it does not seem to be feasible to learn it and its general principles that span across applications in the abstract, from a textbook: you learn it on the job, by apprenticeship, and by trial and error. And if you try to teach it through examples, using actual data, you have to walk a narrow line between getting bogged down in details of the domain-specific background, or presenting unrealistic, sanitized versions of the data and of the associated problems.

Very recently, the challenge posed by these contradictions has been addressed in a stimulating workshop discussion by Sedransk et al. (2010, p. 49), as Challenge #5 – *To use available data to advance education in statistics*. The discussants point out that a certain geological data base "has created an unforeseen enthusiasm among geology students for data analysis with the relatively simple internal statistical methodology." That is, to use my terminology, the appetite of those geology students for explor-ing their data had been whetted by a simple-minded decision support system, see Section 2.5.9. The discussants wonder whether "this taste of statistics has also cre-ated a hunger for [...] more advanced statistical methods." They hope that "utilizing these large scientific databases in statistics classes allows *primary* investigation of interdisciplinary questions and application of exploratory, high-dimensional and/or other advanced statistical methods by going beyond textbook data sets." I agree in principle, but my own expectations are much less sanguine. No doubt, the appetite grows with the eating (cf. Section 2.5.9), but you can spoil it by offering too much sophisticated and exotic food. It is important to leave some residual hunger! Instead of fostering creativity, you may stifle ingenuity and free improvisation by overwhelming the user with advanced methods. The geologists were in a privileged position: the geophysicists have a long-standing, ongoing strong relation with probability and statistics – just think of Sir Harold Jeffreys! – and the students were motivated by the data. The (unsurmountable?) problem with statistics courses is that it is difficult to motivate statistics students to immerse themselves into the subject matter underlying those large scientific data bases.

But still, the best way to convey the principles, rather than the mere techniques of data analysis, and to prepare the general mental framework, appears to be through anecdotes and case studies, and I shall try to walk this way. There are more than enough textbooks and articles explaining specific statistical techniques. There are not enough texts concerned with issues of overall strategy and tactics, with pitfalls, and with statistical methods (mostly graphical) geared toward providing insight rather than quantifiable results. So I shall concentrate on those, to the detriment of the coverage of specific techniques. My principal aim is to distill the most important lessons I have learned from half a century of involvement with data analysis, in the hope to lay the groundwork for a future theory. Indeed, originally I had been tempted to give this book the ambitious programmatic title: *Prolegomena to the Theory and Practice of Data Analysis.*

Some comments on Tukey's paper and some speculations on the path of statistics may be appropriate.

1.1 TUKEY'S 1962 PAPER

Half a century ago, Tukey in an ultimately enormously influential paper (Tukey 1962) redefined our subject, see Mallows (2006) for a retrospective review. It introduced the term "data analysis" as a name for what applied statisticians do, differentiating this from formal statistical inference. But actually, as Tukey admitted, he "stretched the term beyond its philology" to such an extent that it comprised all of statistics. The influence of Tukey's paper was not immediately recognized. Even for me, who had been exposed to data analysis early on, it took several years until I assimilated its import and recognized that a separation of "statistics" and "data analysis" was harmful to both.

Tukey opened his paper with the words:

> For a long time I have thought that I was a statistician, interested in inferences from the particular to the general. But as I have watched mathematical statistics evolve, I have had cause to wonder and to doubt. And when I have pondered about why such techniques as the spectrum analysis of time series have proved so useful, it has become clear that their "dealing with fluctuations" aspects are, in many circumstances, of lesser importance than the aspects that would already have been required to deal effectively with the simpler case of very extensive data, where fluctuations would no longer be a problem. All in all, I have come to feel that my central interest is in data analysis, which I take to include, among other things: procedures for analyzing data, techniques for interpreting the results of such procedures, ways of planning the gathering of data to make analysis easier, more precise or more accurate, and all the machinery and results of (mathematical) statistics which apply to analyzing data.

> Large parts of data analysis are inferential in the sample-to-population sense, but these are only parts, not the whole. Large parts of data analysis are incisive, laying bare indications which we could not perceive by simple and direct examination of the raw data, but these too are parts, not the whole. Some parts of data analysis [...] are allocation, in the sense that they guide us in the distribution of effort [...]. Data analysis is a larger and more varied field than inference, or incisive procedures, or allocation.

A little later, Tukey emphasized:

> Data analysis, and the parts of statistics which adhere to it, must then take on the characteristics of a science rather than those of mathematics, specifically:

> (1) Data analysis must seek for scope and usefulness rather than security.

> (2) Data analysis must be willing to err moderately often in order that inadequate evidence shall more often *suggest* the right answer.

> (3) Data analysis must use mathematical argument and mathematical results as bases for judgment rather than as bases for proofs or stamps of validity.

A few pages later he is even more explicit: "In data analysis we must look to a very heavy emphasis on judgment." He elaborates that at least three different sorts or sources of judgment are likely to be involved in almost every instance: judgment based on subject matter experience, on a broad experience how particular techniques have worked out in a variety of fields of application, and judgment based on abstract results, whether obtained by mathematical proofs or empirical sampling.

In my opinion the main, revolutionary influence of Tukey's paper indeed was that he shifted the primacy of statistical thought from mathematical rigor and optimality proofs to judgment. This was an astounding shift of emphasis, not only for the time (the early 1960s), but also for the journal in which his paper was published, and last but not least, in regard of Tukey's background – he had written a Ph.D. thesis in pure mathematics, and one variant of the axiom of choice had been named after him.

Another remark of Tukey also deserves to be emphasized: "Large parts of data analysis are inferential in the sample-to-population sense, but these are only parts, not the whole." As of today, too many statisticians still seem to cling to the traditional view that statistics *is* inference from samples to populations (or: virtual populations). Such a view may serve to separate mathematical statistics from probability theory, but is much too exclusive otherwise.

1.2 THE PATH OF STATISTICS

This section describes my impression of how the state of our subject has developed in the five decades since Tukey's paper. I begin with quotes lifted from conferences on future directions for statistics. As a rule, the speakers expressed concern about the sterility of academic statistics and recommended to get renewed input from applications. I am quoting two of the more colorful contributions. G. A. Barnard said at the Madison conference on the Future of Statistics (Watts, ed. (1968)):

> Most theories of inference tend to stifle thinking about ingenuity and may indeed tend to stifle ingenuity itself. Recognition of this is one expression of the attitude conveyed by some of our brethren who are more empirical than thou and are always saying, 'Look at the data.' That is, their message seems to be, in part, 'Break away from stereotyped theories that tend to stifle ingenious insights and do something else.'

And H. Robbins said at the Edmonton conference on Directions for Mathematical Statistics (Ghurye, ed. (1975)):

> An intense preoccupation with the latest technical minutiae, and indifference to the social and intellectual forces of tradition and revolutionary change, combine to produce the Mandarinism that some would now say already characterizes academic statistical theory and is most likely to describe its immediate future.
>
> [... T]he statisticians of the past came into the subject from other fields – astronomy, pure mathematics, genetics, agronomy, economics etc. – and created their statistical methodology with a background of training in a specific scientific discipline and a feeling for its current needs. [...]
>
> So for the future I recommend that we work on interesting problems [and] avoid dogmatism.

At the Edmonton conference, my own diagnosis of the situation had been that too many of the activities in mathematical statistics belonged to the later stages of what I called 'Phase Three':

> In statistics as well as in any other field of applied mathematics (taken in the wide sense), one can usually distinguish (at least) three phases in the development of a problem. In Phase One, there is a vague awareness of an area of open problems, one develops *ad hoc* solutions to poorly posed questions, and one gropes for the proper concepts. In Phase Two, the 'right' concepts are found, and a viable and convincing theoretical (and therefore mathematical) treatment is put together.

In Phase Three, the theory begins to have a life of its own, its consequences are developed further and further, and its boundaries of validity are explored by leading it *ad absurdum*; in short, it is squeezed dry.

A few years later, in a paper entitled "Data Analysis: in Search of an Identity" (Huber 1985a), I tried to identify the then current state of our subject. I speculated that statistics is evolving, in the literal sense of that word, along a widening spiral. After a while the focus of concern returns, although in a different track, to an earlier stage of the development and takes a fresh look at business left unfinished during the last turn (see Exhibit 1.1).

During much of the 19th century, from Playfair to Karl Pearson, descriptive statistics, statistical graphics and population statistics had flourished. The Student-Fisher-Neyman-Egon Pearson-Wald phase of statistics (roughly 1920-1960) can be considered a reaction to that period. It stressed those features in which its predecessor had been deficient and paid special attention to small sample statistics, to mathematical rigor, to efficiency and other optimality properties, and coincidentally, to asymptotics (because few finite sample problems allow closed form solutions).

I expressed the view that we had entered a new developmental phase. People would usually associate this phase with the computer, which without doubt was an important driving force, but there was more to it, namely another go-around at the features that had been in fashion a century earlier but had been neglected by 20th century mathematical statistics, this time under the banner of data analysis. Quite naturally, because this was a strong reaction to a great period, one sometimes would go overboard and evoke the false impression that probability models were banned from exploratory data analysis.

There were two hostile camps, the mathematical statisticians and the exploratory data analysts (I felt at home in both). I still remember an occasion in the late 1980s, when I lectured on high-interaction graphics and exploratory data analysis and a prominent person rose and commented in shocked tones whether I was aware that what I was doing amounted to descriptive statistics, and whether I really meant it!

Still another few years later I elaborated my speculations on the path of statistics (Huber 1997a). In the meantime I had come to the conclusion that my analysis would have to be amended in two respects: first, the focus of attention only in part moves along the widening spiral. A large part of the population of statisticians remains caught in holding patterns corresponding to an earlier paradigm, presumably one imprinted on their minds at a time when they had been doing their thesis work, and too many members of the respective groups are unable to see beyond the edge of

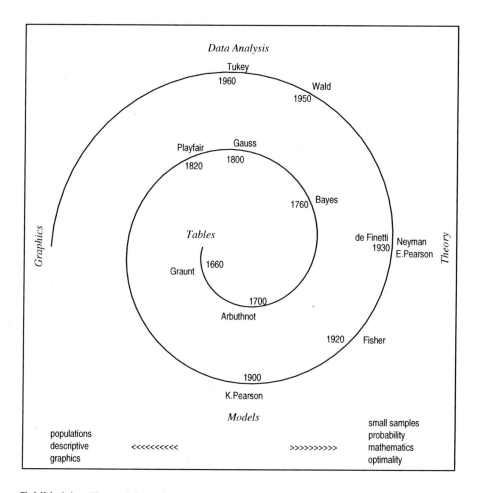

Exhibit 1.1   The evolving spiral path of statistics. Adapted from Huber (1997a),
©Princeton University Press.

the eddy they are caught in, thereby losing sight of the whole, of a dynamically
evolving discipline. Unwittingly, a symptomatic example had been furnished in
1988 by the editors of *Statistical Science*. Then they re-published Harold Hotelling's
1940 Presidential Address on "The Teaching of Statistics", but forgot all about Deming's brief (one and a half pages), polite but poignant discussion. This discussion
is astonishing. It puts the finger on deficiencies of Hotelling's otherwise excellent
and balanced outline, it presages Deming's future role in quality control, and it also
anticipates several of the sentiments voiced by Tukey more than twenty years later.
Deming endorses Hotelling's recommendations but says that he takes it "that they

are not supposed to embody all that there is in the teaching of statistics, because there are many other neglected phases that ought to be stressed." In particular, he points out Hotelling's neglect of simple graphical tools, and that he ignores problems arising from inhomogeneity. It suffices to quote three specific recommendations from Deming's discussion: "The modern student, and too often his teacher, overlook the fact that such a simple thing as a scatter diagram is a more important tool of prediction than the correlation coefficient, especially if the points are labeled so as to distinguish the different sources of the data." "Students are not usually admonished against grouping data from heterogeneous sources." "Above all, a statistician must be a scientist." – Clearly, by 1988 a majority of the academic statistical community still was captivated by the Neyman-Pearson and de Finetti paradigms, and the situation had not changed much by 1997.

Second, most data analysis is done by non-statisticians, and the commonality is hidden behind a diversity of languages. As a consequence, many applied fields have developed their own versions of statistics, together with their own, specialized journals. The spiral has split into many, roughly parallel paths and eddies. This leads to a Balkanization of statistics, and to the rise of sects and local gurus propagating "better" techniques which really are worse. For example, the crystallographers, typically confronted with somewhat long-tailed data distributions, abandoned mean absolute deviations in favor of root mean square deviations some years after the lack of robustness of the latter had become generally known among professional statisticians.

Where are we now, another 13 years later? I am pleased to note that an increasing number of first rate statisticians of the younger generation are committing "interdisciplinary adultery" (an expression jokingly used by Ronald Pyke in his "Ten Commandments" at the Edmonton conference) by getting deeply involved with interesting projects in the applied sciences. Also, it seems to me that some of the captivating eddies finally begin to fade. But new ones emerge in the data analysis area. I am referring for instance to the uncritical creation and propagation of poorly tested, hyped-up novel algorithms – this may have been acceptable in an early 'groping' phase, but by now, one would have expected some methodological consolidation and more self-control.

Though, prominent eddies of the classical variety are still there. I regretfully agree with the late Leo Breiman's (2004) sharp criticism of the 2002 NSF Report on the Future of Statistics ("the report is a step into the past and not into the future"). See also Colin Mallows (2006). In my opinion the NSF report is to be faulted because it looked inward instead of outward, and that prevented it from looking beyond the theoretical "core" of statistics. Thereby, it was harking back to the time *before* Tukey's seminal 1962 paper, rather than providing pointers to the time *after* Tukey's "Future". It gave a fine description of ivory tower theoretical statistics, but it pointedly excluded analysis of actual data. This is like giving a definition of physics that excludes experimental physics. Each of the two has no life without the other.

Despite such throwbacks I optimistically believe that in the 25 years since I first drew the spiral of Exhibit 1.1 we have moved somewhat further along as a profession and now are reaching a point alongside of Karl Pearson, a century later. The focus of concern is again on models, but this time on dealing with highly complex ones in the life sciences and elsewhere, and on assessing goodness-of-fit with the help of simulation. While many applied statisticians still seem to live exclusively within the classical framework of tests and confidence levels, it becomes increasingly obvious that one has to go beyond mere tests of goodness-of-fit. One must take seriously the admonition by McCullagh and Nelder (1983, p.6) "that *all models are wrong*; some, though, are better than others and we can search for the better ones." Apart from searching for better models, we must also learn when to stop the search, that is: we must address questions of *model adequacy*. This may be foremost among the unfinished business. Incidentally, models are one of the two issues where Tukey's foresight had failed (in his 1962 paper he had eschewed modeling, and he had under-estimated the impact of the computer).

But where are we going from here? I do not know. If my spiral gives any indication, we should return to business left unfinished during the last turnaround in the Student-Fisher-Neyman-Egon Pearson-Wald phase. The focus will not return to small samples and the concomitant asymptotics – this is finished business. But we might be headed towards another period of renewed interest in the theoretical (but this time not necessarily mathematical) penetration of our subject, with emphasis on the data analytic part.

CHAPTER 2

STRATEGY ISSUES IN DATA ANALYSIS

This chapter is based on a talk given at the *Conference on Statistical Science Honouring the Bicentennial of Stefano Franscini's Birth*, in Ascona, Switzerland, November 18-20, 1996, see Huber (1997b)[1].

2.1 STRATEGY IN DATA ANALYSIS

> ... *Tactics teaches the use of armed forces in the engagement; strategy, the use of engagements for the object of the war.* (Clausewitz, p. 128)

> ... *die Taktik [ist] die Lehre vom Gebrauch der Streitkräfte im Gefecht, die Strategie die Lehre vom Gebrauch der Gefechte zum Zweck des Krieges.* (Clausewitz, p. 271)

The word "strategy" – literally: "the leading of the army" – inevitably evokes military associations. One is reminded of two famous classics: Sun Tzu's astute and realistic "Art of War", written around 500 BC (here quoted from Griffith's 1963

---

[1] ©With kind permission from Springer Science+Business Media.

translation), and Clausewitz' comprehensive treatise "Vom Kriege", published in 1832 after the author's death (here quoted from the 1984 English translation by Howard and Paret[2], and from the 19th German edition of 1991; I am quoting both versions since the translation by Howard and Paret, while accurate, sometimes misses a precise, untranslatable nuance intended by Clausewitz). In data analysis, strategy is a relatively recent innovation. In the 1980s, in a talk on "Environments for supporting statistical strategy" I had quipped that it was difficult to support something which did not exist (Huber 1986b). Today, the joke might no longer be appropriate, but we still are far away from a Clausewitz for data analysis.

Statisticians have begun to apply the terms tactics and strategy freely and frequently, as evidenced by titles such as "Data Analysis: Strategies and Tactics" (Ehrenberg 1996). But the words are used indiscriminately, and just as in the time before Clausewitz, nobody seems to bother about their precise meaning.

According to Clausewitz, tactics is for battles, while strategy is for campaigns. The standard textbooks of statistics and data analysis are concentrating on techniques geared towards small and homogeneous data sets. I would say: they are concerned with the "tactics" of the field, while "strategy" deals with broader issues and with the question of when to use which technique. In the traditional canon of subjects taught in a statistics curriculum, only (the often neglected) design of experiments and of surveys tend to peek beyond tactical aspects.

The need for strategic thinking (in the sense just explained) is imposed on us by the advent of ever larger data sets. Though, size alone is not decisive. What really forces the issue is that larger data sets almost invariably are composite: they are less homogeneous and have more complex internal structure. Actually, among the four possible combinations, only two seem to occur in practice: small and homogeneous vs. large and inhomogeneous sets. With small and inhomogeneous data one cannot do much statistics. On the other hand, collecting large, homogeneous samples usually is a waste of effort – smaller samples would do the job just as well. Indeed, it is remarkably difficult to find large homogeneous samples matching the ubiquitous "i.i.d. random variables" of theoretical statistics – for example, when I collected materials for my robustness book, the largest homogeneous actual data samples I could find contained fewer than 9000 measurements (Huber 1981, p. 91; Huber and Ronchetti 2009, p. 89).

Strategy is the art of dealing with complex situations and of combining tactical efforts in a larger context. Already in 1940, Deming had admonished the statistical

---

[2] ©With kind permission from Princeton University Press.

profession that as a whole it was paying much too little attention to the need for dealing with heterogeneous data and with data that arise from conditions not in statistical control (randomness). His plea remained largely ignored, even after it had been eloquently reiterated by Tukey in 1962. Curiously, the advent of computers did not bring remedy. Both the collection and the analysis of ever larger data sets became feasible only thanks to computers, and then computer packages took over much of the tactical use of statistical tools. In principle, this should have freed the mind of the statistician of enough drudgery to permit him or her to devote time to strategy. In fact, it did little of the kind, it merely eliminated the informal, subconscious checks performed by a human analyst working with pencil and paper.

Thus, while the writing has been on the wall for over half a century, the traditional teaching of statistics still lags far behind the realities surrounding us. The problems caused by heterogeneous and highly structured data are difficult; I think they have been eschewed precisely because they go beyond tactics and require strategic thinking. Moreover, these problems cannot be harnessed through mathematical formalism, not even the theoretical ones among them. Incidentally, with Tukey (1962) I shall prefer "data analysis" over the synonymous term "statistics" because the latter is used by many in an overly narrow sense, covering only those aspects of the field that can be captured through mathematics and probability.

Clearly, the analogy between military and data analytic strategy is not complete. The main and fundamental difference is that the Data Analyst's adversary (let's call her "Nature") takes on a purely passive stance. The situation is exactly the same as with statistical decision theory and game theory: Nature may be a hard and tricky adversary, but she is not mean. She does not make any conscientious efforts to raise a defense, and she never will go on a counter offensive. As a consequence, the time element – which is so important in military strategy – plays only a secondary role. While blunders can be costly, they rarely are irrevocable and fatal. Very often, the data analyst can recover from them by repeating the analysis, possibly also the data collection, in improved form. Otherwise, the military metaphor is remarkably fruitful.

2.2 PHILOSOPHICAL ISSUES

There are surprisingly close similarities between the philosophical views on the art of warfare and the art of data analysis, and we shall try to exploit some of the analogies. The relevant chapters of Clausewitz furnish ample food for thought, going far beyond the few selected items to be discussed in the following subsections.

2.2.1   On the theory of data analysis and its teaching

Formerly, the terms "art of war" or "science of war" were used to designate only the total body of knowledge and skill that was concerned with material factors. (Clausewitz, p. 133)

Zuerst verstand man unter Kriegskunst nur die Zubereitung der Streitkräfte. (Clausewitz, p. 279)

... The actual conduct of war – the free use of the given means, appropriate to each individual occasion – was not considered a suitable subject for theory, but one that had to be left to natural preference. (Ibid., p. 134)

... Das eigentliche Kriegführen, der freie, d.h. den individuellen Bedürfnissen angepasste Gebrauch der zubereiteten Mittel, glaubte man, könne kein Gegenstand der Theorie sein, sondern dies müsste allein den natürlichen Anlagen überlassen bleiben. (Ibid., p. 280)

... As in the science concerning preparation for war, [the theorists] wanted to reach a set of sure and positive conclusions, and for that reason considered only factors that could be mathematically calculated. (Ibid., p. 134)

... Die Theorienschreiber [wollten], wie in den Wissenschaften der Kriegsvorbereitung, auf lauter gewisse und positive Resultate kommen und also auch nur das in Betrachtung ziehen, was einer Berechnung unterworfen werden konnte. (Ibid., p. 281)

... All these attempts are objectionable. (Ibid., p. 136)

... Alle diese Versuche sind verwerflich. (Ibid., p. 283)

The negative judgments of Clausewitz, such as his disdainful comments on the "Theorienschreiber" (writers of theory) of his time, are not less illuminating for us than his positive contributions. Conventionally, theoretical statistics and data analysis have been concerned only with inventing and refining tools – tests and estimates on one end of the range, procedures and computer programs on the other end. The act of doing data analysis, that is, the free use of those tools in the analysis of actual data, was not considered a suitable object for theory. There are no textbooks; it has to be learned through apprenticeship and anecdotes rather than through systematic exposition. This may very well be the best way to teach it; George Box once pointedly remarked that you do not learn to swim from books and lectures on the theory of buoyancy (Box 1990).

The prevailing opinion has been, and I believe still is, that the global (i.e. strategic) issues in statistics are too elusive to be captured by any theory – perhaps in the mistaken belief that theory has to be mathematical. But it does not follow from this that they should be left entirely to an even vaguer notion of innate statistical common sense, supposedly present in a good data analyst. One is reminded of the contemptuous commentary of Clausewitz on the "writers of theory" of his time.

The need to provide computer assistance to the data analyst now is forcing us to think more systematically about those issues. Their admitted elusiveness however should serve as a warning: any attempt at rigidly harnessing those strategic issues will either miss or kill them. What we need primarily is a suitable *framework* for thinking about them. The situation is parallel to the one prevailing in the military metaphor. Also there, the art of warfare has to be learned by word of mouth and from case studies and ultimately be tested by trial and error in the field. Works on the theory of military strategy, such as the classic by Clausewitz, provide a framework for ordered thinking. There cannot be pedantic textbooks teaching a unique "correct" or "best" behavior. If the military schools try to teach an *unité de doctrine*, the main purpose is to achieve a predictable, consistent behavior between subordinate military commanders. It is a prerogative of the high command to deviate from it, or to order deviations from it. The analogies arising in large data analytic projects are obvious.

2.2.2   Science and data analysis

War is merely the continuation of policy by other means. ... The political object is the goal, war is the means of reaching it, and means can never be considered in isolation from their purpose. (Clausewitz, p. 87)	*Der Krieg ist eine blosse Fortsetzung der Politik mit anderen Mitteln. ... Die politische Absicht ist der Zweck, der Krieg ist das Mittel, und niemals kann das Mittel ohne Zweck gedacht werden.* (Clausewitz, p. 210)

If war is a continuation of politics, then data analysis is a continuation of science. Here, "science" stands as a collective term for all intellectual endeavors concerned with gaining insight in any applied field in the real world. The goals of the scientific project have the primacy, and the data analyst must keep them in mind all of the time. Ask yourself whether the problem is relevant, and whether the data at hand are relevant to the problem. It is far better to give an approximate answer to the right question, which is often vague, than an exact answer to the wrong question, which can always be made precise (Tukey 1962, p. 13f.).

2.2.3 Economy of forces

In life people have first to be taught "Concentrate on essentials." This is no doubt the first step out of confusion and fatuity; but it is only the first step. The second stage in war is a general harmony of war effort by making everything fit together, and every scrap of fighting strength plays its full part all the time. (Churchill, V, p. 365)

[If forces are kept idle,] then these forces are being managed uneconomically. In this sense they are being wasted, which is even worse than using them inappropriately. (Clausewitz, p. 213)

[Es gibt] eine Verschwendung der Kräfte, die selbst schlimmer ist als ihre unzweckmässige Verwendung. (Clausewitz, p. 401)

Athletics [is] one thing and strategy another. (Churchill admonishing Montgomery; Churchill, V, p. 380)

Data analysis is hard and often tedious work, so do not waste forces. Concentrate on essentials! Set priorities (but do not adhere to them slavishly). Do not over-analyze your data. Remember the KISS principle: Keep It Simple and Stupid. Do not demonstrate your data analytic prowess by showing off with the newest and most exotic procedures; instead use the simplest approach that will do the job. Try to convince the client, not to overwhelm him. With most data sets, it is easy to get side-tracked into irrelevancies. In the data analytic literature far too much space and time has been wasted on silly problems and on non-answers to non-questions.

Last but not least, you must know when to stop an analysis, be it whether you have answered the question, or you have squeezed the data dry. If it comes to the worst, and it is impossible to answer the question on the basis of the available data, you ought to concede this fact and accept defeat gracefully, even though such a capitulation can be extremely painful.

2.3   ISSUES OF SIZE

*In war, numbers alone confer no advantage. Do not advance
relying on sheer military power.* (Sun Tzu, p. 122)

*Generally, management of many is the same as management of
few. It is all a matter of organization.* (Sun Tzu, p. 90)

*In this connection it appears appropriate to recall the Emperor
of China story attributed to the astronomer Kapteyn, for which
I am indebted to Prof. G. Uhlenbeck. The exact height of the
Emperor could be obtained by asking each of the 500,000,000
Chinese to guess at his height. It was not necessary for any
of his subjects to have seen him, or even his picture, because
the application of statistical methods to so many 'individual
observations' would give an answer for the Emperor's height
to a precision of a few microns...! It is clear that millions of
measurements of say a table with a meter stick will not give an
average measurement accurate to an Ångström unit.* (W. Parrish
1960, p. 849)

Also in data analysis, numbers alone confer no advantage. Low data quality
cannot be counterbalanced by data size – Kapteyn's parable illustrates this fact in
an inimitable fashion. What is at issue here is that with actual data it is only rarely
possible to overcome a poor signal-to-noise ratio by increasing the data size: if the
random errors are large, then the systematic errors usually are too, and the latter may
negate the hoped-for advantage of the larger data size.

But there are cases where large data volume is an intrinsic necessity, and we clearly
need to think about strategies for dealing with very large data collections.

With increasing data sizes, in particular with data that must be stitched together
from multiple sources (e.g. multiple hospitals), organizational problems of data base
management and quality control can grow to become a headache.

Since the advent of computers and their employment for data collection we are
witnessing an explosion of data sizes. Already the paper by Butler and Quarrie (1996)
about high energy physics experiments was concerned with data in the petabyte (10^{15}
bytes) range, this being the mere remainder after cutting down a much heavier data
flow by preprocessing.

In a paper on analysis of huge data sets (1994b), I had proposed a classification of data sets by size in steps of a factor 100 from "tiny" (10^2 bytes), "small" (10^4 bytes), "medium" (10^6 bytes), "large" (10^8 bytes) to "huge" (10^{10} bytes) and "monster" (10^{12} bytes). Note that upscaling by factor of 100 is highly non-trivial, since it causes profound qualitative changes: one needs not only different media for data storage, but the data typically have very different, more complex characteristics.

Originally, I had omitted terabyte monster sets, because I had felt such sizes to be unmanageable. Wegman (1995) only underscored my misgivings when he gave that size the name "ridiculous". The issue here is: up to what size can we have a meaningful *data analysis* in the strict sense (requiring judgment based on information contained in the data, that is requiring human inspection), as opposed to *data processing* (automated, not requiring such judgment)? Wegman pointed out that the human eye cannot resolve more than a few megapixels; it follows that exhaustive human inspection of the raw data will become infeasible beyond "large" sets. See also Chapter 3.

"Huge" and larger sets must be preprocessed down to manageable sizes before we can begin an analysis in the above sense. Though, such preprocessing may prejudge the analysis by suppressing important, but unanticipated features.

Datasets in the tera- to petabyte range (10^{12} to 10^{15} bytes) are a real headache – if one is scaling by a factor 100 or more beyond actual past experiences, one inevitably encounters qualitatively new, unanticipated aspects. In the absence of well understood strategies for overcoming such obstacles, a buzz word has been invented as a placeholder: Data mining. The idea is to somehow program a computer to grind through a huge pile of data in a patient and intelligent fashion, in the hope that it will eventually find valuable structural nuggets. The idea clearly is intriguing, but unfortunately it is very elusive and subject to hype.

Statisticians have reacted very negatively to utterly naive enthusiastic descriptions of data mining that gave the unfortunate impression that it was a mere collection of all known methods of applied statistics, used uncritically. See Fayyad et al. (1996). p. 22, for cautionary remarks by proponents. I am inclined to view data mining not as referring to any specific techniques, but teleologically, as comprising all methods devised to reduce very large data volumes down to a size fit for human inspection and analysis, by any means whatsoever. Despite the advances of the past years, I do not think that I am doing injustice to the prevalent situating by contending that a very large part of what is going under the label of data mining continues to be a nearly empty hull, held in place by hot air, at best serving as a place-holder for more substantive contents to come.

Naive data mining, that is: grinding a pile of data in a semi-automatic, untutored fashion through a black box, almost inevitably will run into the GIGO-syndrome – Garbage In, Garbage Out. Unfortunately, you may not recognize garbage output as such. The data mining community does not yet seem to be sensitized to the specific problems arising with statistical data, where relationships hold only on average, and where the average can be distorted by selection bias or similar effects. The more opaque a black "data mining" box is, the less likely it is that one will recognize potential problems. A case story from a data analysis exam may serve as a warning. The task was to distinguish between carriers and non-carriers of a certain genetic disease on the basis of enzyme and other data. A student found that age was the variable that discriminated best between carriers and non-carriers. On the face of it, this was entirely correct, but useless. What he had discovered, but misinterpreted, was that in the process of data collection, carriers and controls had not been properly matched with regard to age. Would you have spotted the problem if the result had been presented not verbally but in the form of a duly cross-validated black box (e.g. a neural network)?

Though, there have been some convincing success stories of "non-naive" data mining (cf. Chapter VII of Fayyad et al. (1996)). To my knowledge, the earliest success of sophisticated data mining is the surprising discovery of the Lunar Mascons by Muller and Sjogren (1968), who extracted structure from the residual noise of a Doppler radar ranging experiment. They did this work on their own time, because their superiors at JPL (Jet Propulsion Laboratory) felt they were going to waste their efforts on garbage. I remember Muller joking in the early 1970s that evidently one person's junk pile was another's gold mine, so not only the invention of data mining, but also giving it its colorful name ought to be credited to them. Their success was due not to a black box approach, but to a combination of several thoughtful actions: First, a careful inspection of residuals, revealing that these did not behave like white noise. Second, a tentative causal interpretation: the cause might be an irregular distribution of lunar mass. Third, modeling this distribution by expanding it into spherical harmonics with unknown coefficients and then estimating these coefficients by least squares. Fourth, a graphical comparison of isodensity contours of the estimated mass distribution with a map of lunar surface features. The discovery literally happened at the moment when the plot emerged from the plotter: it revealed systematic mass concentrations, or Mascons, beneath the lunar *maria*, see Exhibit 2.1. Interestingly, the persuasive argument in favor of correctness of the result was not probabilistic (i.e. a significance level, or the like), but the convincing agreement between calculated mass concentrations and visible surface features.

Exhibit 2.1   Lunar Mascons. Mass isodensity contours plotted on top of a map of lunar surface features. From *Science* Vol. 161, No. 3842, 16 August 1968. Reprinted with permission from AAAS.

2.4   STRATEGIC PLANNING

Data analysis ranges from planning the data collection to presenting the conclusions of the analysis. The war may be lost already in the planning stage. It is essential never to lose sight of the overall purpose of the endeavor, to balance the efforts, to identify decisive details, and to pay attention to them. The following aspects deserve early attention.

2.4.1   Planning the data collection

Unfortunately, the data analyst rarely has any control at all over the earliest phases of an analysis, namely over planning and design of the data collection, as well as the act of collecting the data. The situation is not improved by the fact that far too few statisticians are able and willing to take a holistic view of data analysis.

Data can be experimental (from a designed experiment), observational (with little or no control over the process generating the data), or opportunistic (the data have been collected for an unrelated purpose). Massive data sets rarely belong to the first category, since by a clever design the data flow often can be reduced to manageable proportions already before it is recorded. They often belong to the third category for plain reasons of economy: it is cheaper to recycle old data than to collect new ones.

The principal questions to be raised and addressed thus are: What is the purpose of the analysis? Do we already have in hand suitable data? How can we collect such data and in such a way that we will have a reasonable chance to achieve our purpose? Will the data contain the information we are looking for and which is necessary for answering our questions? Will the questions still be the same by the time the data will be in hand? There are cases (mostly hushed up and therefore rarely documented) where multimillion dollar data collections had to be junked because of their poor design. An example is the predecessor of the hail-prevention experiment described by Schmid (1967). The story was told to me by Jerzy Neyman – Schmid does not mention it. Schmid must have been an unusually resourceful and persuasive statistician, since he was able to convince the sponsor that the data collected in the previous experiment and which he was supposed to analyze were worthless because one had neglected to randomize the experiment.

You should plan the data collection with the subsequent analysis in mind, clever planning may simplify the analysis. This applies in particular to questionnaire design. Match the prospective analysis methods to data size. Computer intensive methods do not work with very large data – procedures with computational complexity above

$O(n^{1.5})$ simply take too much time. The problem will not go away with the growth of computer performance since data sizes will grow too. Such considerations may eliminate, for example, all standard clustering approaches, since they exceed the just-mentioned complexity limit.

Already in the planning stage, strategic thinking about error control measures is mandatory: how should we plan the collection and the coding of the data in order to make spotting and correcting errors easier? One should plan to provide some redundancy, it will help with error detection and correction. Data checking is never quite finished, unrecognized errors may surface at any stage of the analysis. Highly optimized sampling designs typically leave no degrees of freedom to detect and measure errors in the model. Moreover, with such designs, one often is lost if the questions to be answered are even slightly shifted in the course of the project, or if one wants to re-use that data for another purpose.

The meta-data, that is, the story behind the data, why and how they were collected, how they were pre-processed, the meaning and the names of the variables, and so on, are just as important as the data themselves. Curiously, meta-data tend to get separated from the data early and are subsequently lost. While data sets are electronically mailed around the globe, meta-data typically remain committed to obscure technical reports and sometimes to oral tradition only.

2.4.2 Choice of data and methods.

> *Strategy fixes the point* where, *the time* when, *and the forces* with *which the battle is to be fought.* (Clausewitz)

> *Die Strategie bestimmt den Punkt,* auf *welchem, die Zeit* in *welcher, und die Streitkräfte,* mit *welchen gefochten werden soll.* (Clausewitz, p.373)

In data analysis, a crucial and often bungled decision is the choice of data to be investigated. It may pay to ignore a massive majority, if it is distorted by systematic errors.

Equally crucial is the choice of methods. Very often, a statistician or scientist chooses a method because he or she is familiar with it, rather than because it is appropriate. I know of cases where the data cried for a survival analysis approach but were treated by regression analysis.

2.4.3 Systematic and random errors

Quantity never compensates for quality; the quality of the data base is of paramount importance. Error checking (after the fact) does not compensate for error prevention. Some of our worst experiences have been with distributed data that were routinely collected over many years, by many people, at many places, but which were never really put to use until much later, when it was too late for corrective measures. Without organized early feedback, internal inconsistencies, misinterpretations and just plain errors begin to accumulate. Well-meant "improvements" may spoil comparability between batches collected by different groups of people, or during different time periods. Records kept of highway maintenance may serve to illustrate this point: because of the geographical spread, the extended time frame and the distributed administrative responsibilities, it is difficult to enforce uniform quality control.

When human beings collect and record data, they notoriously commit gross errors, in a more or less random fashion, with an overall frequency between 1% and 10%. Robust methods cope quite well with such errors. But with larger than usual sets, and with more or less automated data collection, new types of gross errors arise, in particular gross errors systematically affecting entire branches in a hierarchical data organization. Some of them may not be errors in the usual sense of the word, perhaps they merely are different interpretations of ambiguous specifications, but they destroy comparability between branches. We need strategies for dealing with gross systematic errors.

Traditional present-day statistical theory recognizes only random errors, and probability models are supposed to be exact, unless shown otherwise by a suitable goodness-of-fit test. If we accept this view, there are no biases except those caused by the use of slightly inappropriate estimates. With ever larger data sets, these issues need to be rethought. Goodness-of-fit tests will reject even perfectly adequate models, once the sample size is large enough, simply because not every nook and cranny can and should be modeled. But apart from that, we must finally recognize that a model is but a simplified and idealized image of reality, abstracting from irrelevant features. We need strategies for dealing with models that are known to be imprecise.

2.4.4   Strategic reserves

A reserve has two distinct purposes. One is to prolong and renew the action; the second, to counter unforeseen threats. The first purpose presupposes the value of the successive use of force, and therefore does not belong to strategy. ... But the need to hold a force in readiness for emergencies may also arise in strategy. Hence there can be such a thing as a strategic reserve, but only when emergencies are conceivable. (Clausewitz, p. 210)

Eine Reserve hat zwei Bestimmungen, die sich wohl voneinander unterscheiden lassen, nämlich: erstens die Verlängerung und Erneuerung des Kampfes, und zweitens, der Gebrauch gegen unvorhergesehene Fälle. Die erste Bestimmung setzt den Nutzen einer sukzessiven Kraftanwendung voraus und kann deshalb in der Strategie nicht vorkommen. ... Das Bedürfnis aber, eine Kraft für unvorhergesehene Fälle bereit zu haben, kann auch in der Strategie vorkommen, und folglich kann es auch strategische Reserve geben: aber nur da, wo unvorhergesehene Fälle denkbar sind. (Clausewitz, p. 397)

In traditional theoretical statistics, there is no place for unforeseen events. Though, practitioners always have known that part of the data should be kept back, because without such reserves one may not be able to deal with surprises. On the basis of the same data, you cannot both discover a feature and perform a meaningful statistical test for its presence.

It is therefore necessary to provide strategic reserves, so that one can test surprises: are they mere statistical flukes or real, novel effects? If in the course of analyzing a batch of data an unanticipated feature is discovered, and if one then devises a statistical test for testing the hypothesis that the feature is a mere random artifact against the alternative that it is real, then, at least in my experience, such tests never yield very large or very small P-values. The putative reason is: if the P-value is larger than, say, 0.1 or 0.2, one would not have noticed the feature, and hence one would not have tested; if it is smaller than, say, 0.001 or 0.01, one would have accepted the feature as real without feeling the need for a test. But given that under the null hypothesis the P-values are uniformly distributed in the interval $(0,1)$, and if we are only testing cases where the P-value is ≤ 0.1, we are performing a conditional test of a special kind. As a consequence the actual level, that is the (conditional) probability of rejecting the true null hypothesis, is an order of magnitude larger than the nominal level: 50% instead of 5%, or 10% instead of 1% (see Huber 1985a).

Admittedly, this example is somewhat extreme. But it shows that it is no longer possible to calculate reliable P-values *after* one has looked at the data – unless they are based on fresh data, they may be worse that useless, namely misleading.

For typical cases where one would not bother to quantify the conclusions by P-values (rightly so!), see the examples on Lunar Mascons in Section 2.3 and on Radon levels in Section 3.8.

2.4.5 Human factors

All too often, the people responsible for the data collection make unwarranted (often tacit) assumptions about the model behind the data or behind the collection process. Unfortunately but typically, the latter then is designed in such a way that the validity of the assumptions cannot even be checked. Example: in the absence of actual measurements, there is no reason to assume that after an intervention (repair, maintenance service) the system has been restored to the ideal "good" state.

There are harmful idiosyncrasies both on the side of the statisticians and of the scientists, and they may cause awkward communication problems. For example, statisticians tend to think in terms of significance levels and of controlled experiments, and reciprocally, scientists expect significance statements from the statisticians. However, if the data do not derive from a designed experiment, then the observational character of the data set by itself may force entirely different strategies upon the analyst. In particular, it may be difficult or impossible to quantify conclusions in probability terms; significance levels expected by the scientist (or by the journal editor!) may be worse than meaningless.

A statistician rarely sees natural science data in their original, raw form. Usually a considerable amount of preprocessing has already been performed. After a while, standard procedures tend to become invisible, especially now, when such procedures are packed into computer programs used as black boxes, and the client will fail to mention them to the statistician. A common problem is that many scientists are aware of only one type of random effects, namely observational errors. Preprocessing performed by them is geared towards reducing the effects of observational errors, both systematic and random; it may distort and falsify the stochastic structure underlying the data. An example occurs in Section 5.7.

Visualization of highly structured data is difficult in any case. With the presentation of large data sets we run against the previously mentioned limitations of the human perceptual system. The comments of Taubes (1996) on "the problem of displaying the output of a petaflops computer to the user, which could mean somehow communicating 100 terabytes of data", miss the point. The problem is not, as it is stated there, "to provide a thick pipe from the supercomputer to the outside world," but on the contrary, to narrow the data flow down to a rate that can be digested by human beings.

2.5   THE STAGES OF DATA ANALYSIS

The following subsections are intended as a kind of check-list to assist with overall planning of the data collection, and with the preparation of resources for the stages of the analysis after the data have been collected. All stages are strategically important; the precise way how their purpose is achieved is irrelevant and a matter of tactics. I shall describe these stages roughly in the order in which they are encountered in typical analyses, and shall try to illustrate them by examples. Though, strictly speaking, ordering the pieces is impossible, one naturally and repeatedly cycles between different actions. The stages in question are:

- Inspection
- Error checking
- Modification
- Comparison
- Modeling and Model fitting
- Simulation
- What-if analyses
- Interpretation
- Presentation of conclusions

2.5.1   Inspection

The principal purpose of inspection is to see things one was not looking for. Inspection occurs on many levels, in many guises. Initially, it helps the analyst to become familiar with a new data set. Most often, its purpose is quality control: preliminary error checking of the raw data, or checking whether an interactive analysis is still on track. The other most frequent use is hunting for unsuspected or unusual features: such as outliers, clusters or curvature. Still another occurs in connection with comparison and modeling: qualitative inspection of the fit between model and data.

Sometimes, a visual inspection of ASCII data can be extremely helpful (is the structure of a file the same from beginning to end?). But ordinarily, inspection is graphical. Facilities for interactive creation and manipulation of subsets can make inspection an extremely powerful tool in various stages of the analysis.

The problem with inspection is that only a very limited amount of information can be meaningfully presented to, and digested by, a human being at any one time. With larger data sets one must rely on subset selection (random or targeted) and summarization (such as averages and density estimates); with huge, highly structured sets, all approaches are guaranteed to leave large, unexplored and unseen holes in the data landscape. It is not easy, but necessary, to identify and to occupy the strategically important positions.

2.5.2   Error checking

If something can go wrong, it will. (Murphy's Law)

Before one even can begin to check a data set for errors, one must be able to read it. Before a data set reaches the analyst, it usually has been copied and recoded a few times, and on the way to the forum some funny things may have happened. A couple of anecdotes will illustrate some frequently encountered problems. For example, a binary file may have been prepared on a big-endian computer and you happen to have little-endian hardware (i.e. the bytes in a computer word are ordered from left to right or from right to left, respectively). More than once in our experience, some automatic, overly clever data transfer program did mistake a binary file for an ASCII file and substituted CR-LF (carriage-return-line-feed) sequences for all bytes looking like a LF character.

Imagine you have been shipped a large data file, in the megabyte to gigabyte range, and somewhere in the middle of the file a few bytes are mutilated, garbled or missing. Can you do the following tasks, how easily can you do them, and how long will it take you to complete them?

- Read the file?

- Read beyond the damaged segment?

- Locate, inspect and repair the damaged segment?

The usual data base programs and data formats perform poorly on these tasks. With some data formats, re-synchronizing the reading program after a mutilated segment may be impossible. After a few encounters with such problems you will long for robust, plain old ASCII files with fixed record lengths, with end-of-line markers (this helps with re-synchronizing the reading program after a trouble spot), and with recognizable separators (such as blanks) between data items. After a short while you will begin to write your own robust data reading and editing programs that can handle large files.

Traditional methods of data checking employ a combination of the following three approaches:

- legality checks: are the values physically or legally possible?

- plausibility checks: are the values likely?

- consistency checks: are the values internally compatible?

The typical actions triggered by a discovered error comprise inspection, perhaps supplemented by a collation of the raw data sheets, followed by an individual decision by the human analyst; automatic correction (replacement by missing values, by imputed values, or by the nearest plausible values); rejection (elimination) of entire cases. However, this catalogue of actions is inadequate, since it is based on the unwarranted and nearly always wrong assumption that the errors are isolated and independent. More often than not they are of a systematic nature. One should look into possible patterns of errors. The following case story shows what can happen.

The Widows-and-Indians Syndrome. A beautiful detective story about a case of automatic data checking that went wrong is described in a classic paper by Coale and Stephan (1962). They accidentally noted that the 1950 U.S. census had found surprisingly many young widowers. Curiously and improbably, there were more 14-year-old widowers than 15-year-old ones, and again more than 16-year-old ones. The subsequent analysis by Coale and Stephan demonstrated convincingly that a few thousand cards must have been punched wrongly, so that part of the data was shifted one character position. Because of this shift, heads of households in their forties became 14-year-old widowers, and if they were in their fifties, they became 15-year-old widowers, and so on. In principle, the checking program could and should have spotted the problem and alerted the analysts through a pattern of discoverable errors: 13-year-old and younger widowers had been thrown out as legally impossible. The identical shift left traces by inflating also some other small categories, in particular one involving American Indians. The Indians clinched the case.

There is a lesson to be learnt from this story: automatic spotting and elimination of gross errors is not good enough. Whenever there is a pattern of gross errors, their cause must be analyzed.

Variants of the syndrome are widespread. Most often, the real problem is that the coding book or the recording hardware provide insufficiently many digits for certain data items, and the problem is either overlooked or the remedies remain undocumented. Some examples from my personal experience include:

(1) In one case, time was measured in seconds since midnight, but erroneously recorded as short integers (which range from 0 to 65535 only, while the day has 86400 seconds). Values above 65535 were reduced modulo 65536, but with some effort the correct values could be reconstructed through monotonicity.

(2) In another case, the coding book had provided one decimal digit for a small integer, and then occasional values 10 and 11 turned up in the data. The keyboard operators cleverly decided to use something like hexadecimal codes for those values. Much later, these codes were trapped as errors by the checking program, and if the person responsible for the analysis had not been unusually circumspect, we never would have realized that legitimate two-digit values did occur in this kind of data, and the conclusions would have been seriously biased.

(3) In a variant of this example, a programmer had managed to squeeze additional information into the unused 8th bit of 7-bit ASCII characters, making the data seem garbled.

(4) In still another case, the coding of a variable was expanded from 2 to 3 digits, but one did not record this fact in the code book, and the wrong field size persisted also in the header part of the data base file.

In short, while actual data coding usually changes and improves over time, its details never agree with the code books – often not even with the headers of the data files themselves! – and attempts to correct coding deficiencies usually remain undocumented and will later baffle the analyst.

In a weird case, an unnecessarily high temporal resolution of the raw data had subsequently been reduced by numerical averaging. As a side effect – presumably because the coding book had not drawn attention to the specific details of the encoding – some important, tightly packed information was irretrievably garbled.

Some of the most insidious errors, slipping through most data checks, are holes in the data and duplicated batches of data – it is easy to skip a box of data sheets, or to enter it twice. Even worse, it sometimes happens that a batch is entered wrongly first, then the error is discovered and corrected, but the corrected version mistakenly is exchanged against a perfectly good segment.

2.5.3   Modification

Data modification ranges from simple transformations, like taking logarithms in order to equalize variances in spectrum estimates, to complex aggregation or grouping operations. Because of the limited dynamic range of the human eye (if features have a size disparity of 1:30, visual comparisons fail), modification often is an indispensable preparation for graphics.

On the strategy level, modification serves primarily for the preparation of a streamlined base set, in which the irregularities and inconsistencies of the raw data have been ironed out, as well as for that of various derived sets (cf. Section 3.5.4).

2.5.4   Comparison

Already Playfair (1821) had stressed the importance of comparison: "It is surprising how little use is derived from a knowledge of facts, when no comparison is drawn between them." Playfair is considered the father of statistical graphics, and it certainly is no accident that the main tools of comparison are graphical.

With statisticians, by a kind of Pavlovian reaction, comparison between model and data typically provokes the calculation of goodness-of-fit tests. In data analysis of moderately sized samples, goodness-of-fit tests often serve with a slightly different twist, namely not as a test of the model (which is known to be inaccurate anyway) but as a signal, warning against the danger of over-interpreting the data. As a consequence of the different twist, goodness-of-fit tests in data analysis typically are (or at least: should be) performed with significance levels well above the customary 5%.

Comparison between results from two or more unrelated sources of information may provide a more convincing confirmation of the correctness of an interpretation than any statistical test (cf. the example of the Lunar Mascons, mentioned in Section 2.3).

Remember that overly complex designs can make comparisons difficult.

2.5.5   Modeling and Model fitting

Modeling is a scientific, not a statistical task. This creates difficulties on the interface both on a human level, between the scientist and the data analyst, and on the software level, between programs of different origins (whose bug is it?).[3] It helps somewhat if the scientist and the statistician are one and the same person putting on two different hats. Modeling notoriously involves a lot of non-trivial *ad hoc* programming, precisely because of interface problems. A data analysis package is useless if it cannot easily be interfaced with existing outside programs for simulation loops and the like, and if it does not facilitate the *ad hoc* programming of such models.

Insight is gained by thinking in models, but reliance on models can prevent insight. The "discomfort" mentioned by Butler and Quarrie (1996), p. 55, in connection with streamlined data collection in high energy physics must be taken very seriously: Automated data screening based on a model may hide all evidence of phenomena that lie outside the model or even contradict it.

Increasingly, models are nonlinear and must be fitted by nonlinear methods. Nonlinear fitting procedures always are tricky and can go wrong; graphical comparison becomes absolutely essential, not only for checking the quality of the fit and of the model, but often also to see which part of the data is relevant for which part of the fit. Eye-balling, that is manually adjusting parameters until the fit is visually acceptable, sometimes is preferable to formal optimization, especially in the early stages of an analysis: an incomplete model may foul up the formal fit in unexpected ways.

2.5.6   Simulation

With more complex data and more complex analysis procedures, simulation gets ever more important, and this not only because theory is no longer able to furnish exact probability values.

Resampling methods (bootstrap) have been overrated. Resampling works best with unstructured, homogeneous data (that is: with data that can be viewed as a sample from a larger population). It still works with well-designed stratified samples. It fails with highly structured data (which are becoming the rule rather than the exception). Simulation still works: prepare synthetic data sets similar to the real one

---

[3]In 1948 Norbert Wiener wrote: "[The physiologist Dr. Arturo] Rosenblueth has always insisted that a proper exploration of these blank spaces on the map of science could only be made by a team of scientists, each a specialist in his own field but each possessing a thoroughly sound and trained acquaintance with the fields of his neighbors; all in the habit of working together, of knowing one another's intellectual customs, and of recognizing the significance of a colleague's new suggestion before it has taken on a full formal expression. The mathematician need not have the skill to conduct a physiological experiment, but he must have the skill to understand one, to criticize one, and to suggest one. The physiologist need not be able to prove a certain mathematical theorem, but he must be able to grasp its physiological significance and to tell the mathematician for what he should look." (Wiener 1963, p. 3)

and check how the analysis performs on them. Admittedly, the results depend on the somewhat arbitrary choice of the model behind the simulation, but a comparison between simulated synthetic data sets and the actual one (usually such comparisons are not trivial) will give an idea of the phenomenological quality of the model, and one can get at least some crude estimates of the variability of estimates.

Data preprocessing and processing get ever more complex, thus there are more possibilities to create artifacts by the processing methods. In data analysis, simulation therefore has a special and important role: often, the only way to recognize artifacts of processing is to generate simulated data and to subject them to the identical processing as the real data.

2.5.7   What-if analyses

The more complex the data structure and the possible explanatory models, the more important it is to conduct alternative what-if analyses. In this category, we have anything from relatively simple sensitivity analyses to involved checks of alternative theories. What happens if we omit a certain subset of the data from the analysis? What if we pool some subsets? What if we use a simpler, or a more complicated, or just a different model?

2.5.8   Interpretation

Interpretation, just like modeling, belongs into the domain of the scientist. Statisticians tend to think of interpretation in terms of "inference", that is, a Bayesian will assign probabilities to statements, a frequentist will think in terms of tests and P-values, scientists expect from the statisticians assistance with the quantification of the conclusions. This is too narrow a framework, covering only that subset of interpretation that can be numerically quantified by probability values (P-values, significance levels, confidence intervals, and so on).

2.5.9   Presentation of conclusions

The larger the data sets are, the more difficult it is to present the conclusions. The presentation must be adapted to the language and customs of the customers, and one may have to educate them – and in particular the journal editors! – that sometimes a P-value is worse than useless (cf. Section 2.4.4). With massive data sets, the sets of conclusions become massive too, and it is simply no longer possible to answer all potentially relevant questions. We found that a kind of sophisticated decision support system (DSS), that is: a customized software system to generate answers to questions of the customers, almost always is a better solution than a thick volume of precomputed tables and graphs. It is straightforward to design a system duplicating the functions of such a volume, and it is easy to go a little beyond, for example by providing hypertext features or facilities for zooming in on graphs. But the appetite grows with the eating, trickier problems will arise, and the DSS then begins to develop into a full-fledged, sophisticated, customized data analysis system adapted to the particular data set(s). See also Section 3.7.6.

Actually, with massive data sets the need for customized data analysis systems arises already earlier in the analysis, namely whenever several people with similar needs must work with the same data, or the same kind of data, over an extended period of time. It is humanly impossible to pre-specify a customized system in advance, one must learn by trial and error. Close cooperation and feedback between data analysts, subject area specialists and end-users are required, and whenever possible, the latter must be involved from the very beginning. Thomas Huber and Matthias Nagel (1996) have described the methodology for preparing such systems under the name "Data Based Prototyping". The process is driven by the data and by its on-going, evolving analysis; it is a task which must be done by people analyzing the data, and it cannot be left to mere programmers.

2.6   TOOLS REQUIRED FOR STRATEGY REASONS

The preceding sections have exposed the need for a number of relatively specific resources or supporting tools that ought to be provided on the strategy level. We have experimented with all of them, and the following statements have been tempered both by trials and by errors. In the following, I shall list these tools more systematically and sketch their respective roles.

2.6.1   Ad hoc programming

We found that *ad hoc* programming is needed on all levels, and that it always involves much trial and error:

- On the lowest level: in the data checking and cleanup stage, for reading and writing arbitrary data representations, binary and otherwise.

- On higher levels: in the data analysis stage for facilitating the analysis itself; for modeling; for simulation; and for making customized data analysis systems.

In order to facilitate *ad hoc* programming, a good programming environment is crucial, including an intuitive command and response language that can serve both as a high-level user interface and as a general purpose programming language (cf. Chapter 4). A properly prepared, carefully chosen collection of basic commands or building blocks is essential, with blocks that are neither too small nor too large, and it must be possible to expand this collection freely by combining blocks into new entities obeying the same syntax and interface rules. Blocks that are too small lead to a kind of tedious assembly programming, blocks that are too large are difficult to combine. The currently fashionable emphasis on mouse driven graphical (and therefore overly rigid) user interfaces is counter-productive, and so is excessive insistence on "re-usable" software. Building blocks are specifically designed for re-use. While data analysis software from previous projects can serve as a welcome collection of templates for new programs, it hardly ever is directly re-usable. The specific needs change too much from one project to the next.

2.6.2 Graphics

On the strategy level, graphics is needed mainly for "intelligence" purposes. This comprises in particular the following tasks:

- For inspection and for detecting unexpected features.

- For various comparisons.

- For internal documentation and planning.

There is an important, sliding distinction between exploration graphics and presentation graphics. They have contradictory requirements with regard to ease and speed of production and of quality. Exploration graphs must be easy and quick to generate, and the user must have an intuitive assortment of high-interaction tools to help with the interpretation of a plot, for example for disentangling congested regions of a scatterplot (like those occurring in some graphs of Chapter 7) by zooming or rotating, for selective labeling, or for creating and temporarily highlighting some subsets through colors and symbols. Often, they are produced *ad hoc*, but they might have a very complex structure, understandable only to their originator and only at the time of their generation. Mostly, exploration graphs are short-lived screen graphs. Presentation graphs must be polished and pared down to explainable essentials. Graphics produced for internal documentation are somewhere in between. For the latter purpose, for example, time-stamps on all graphs are quite important; if such time-stamps appear on presentation graphs, most people seem to resent them.

2.6.3   Record keeping

Record keeping is needed for three main purposes:

- For auditing the correctness of an analysis.

- For repeating an analysis in a slightly modified form: correction of errors, what-if analyses, repetition with a new data set, and so on.

- As a basis for reports, or for the construction of decision support systems.

Records must be kept in a form that can be parsed and amended easily by human users for purposes of re-execution.

In my experience, the main problems with record keeping, or more precisely, with keeping order among records, are caused by a combination of *ad hoc* programming, what-if analyses and alternative simulation. One can produce chaos in no time by accumulating a pile of slightly different analyses. Tables and graphs may have been produced with slightly different programs or parameter values, and some of them may be obsolete or wrong. The last version is not necessarily the best version.

2.6.4   Creating and keeping order

To err is human; to really mess up things you need a computer.
(Ancient Proverb)

With large data analyses, it is strategically vital to create and to preserve order. It is necessary to distribute tasks between human beings and machines in a proper fashion, and to force the human analysts to pay attention and never to let lapse an analysis into a state of chaos by accumulating too much garbage.

While record keeping accumulates information, the creation of order does the opposite: it decreases entropy, that is, it destroys information in a selective fashion. In mythology, creation is a job for gods, who traditionally create worlds out of chaos. It seems that creation of order is too demanding a task to be left to a mere machine.

At best, the machine may help to preserve order. But with increasing age, and I hope, wisdom, I have begun to have second thoughts even about that. Back in 1986 I had formulated postulates for sophisticated "Lab Assistant" systems that would be able to interpret session records and to identify forward and backward dependencies (Huber 1986b). Prototype systems were subsequently programmed by Bill Nugent and by Werner Vach (1987). Clearly, such systems help an analyst with keeping an overall view of a complex and lengthy data analysis, and will assist him or her with the retrieval of crucial items. But will it ever be possible to design a successful Lab Assistant system serving more than one master, namely an entire team of data analysts working with massive data? Such systems will run into problems caused by human factors and in particular by the requirements of communication between humans. While it is essential to keep track of modifications, it is devilishly tricky to do so by machine in such a way that somebody else than the originator of the modifications can make sense of a pile of slightly different and sometimes slightly wrong analyses. A few months later, even the originator will be baffled. See also Section 4.4.4.

The human act of creating subjective order by separating the wheat from the chaff, and throwing out the latter, also creates insight, and no machine can (or should!) replace the creative processes going on in the human mind.

CHAPTER 3

MASSIVE DATA SETS

Prefatory note. *On July 7-8, 1995, a workshop on the statistical analysis and visualization of massive data sets, organized by Jon Kettenring, with more than 50 participants, was conducted at the National Research Council's facilities in Washington, D.C. The proceedings of that workshop were published by Kettenring and Pregibon (1996).[1] My own contribution to the workshop – an attempt to summarize and synthesize the issues – had been subtitled "The Morning After"; it had actually been drafted on the flight back home after the workshop (see Huber 1996a). Clearly, some of my statements are dated, for example tasks that then required a super-workstation now can be handled on standard PCs. My main conclusions remain standing, so I have left the immediacy of my responses intact, except that I have shortened or excised some clearly dated material and have added some italicized afterthoughts benefitting from hindsight.*

[1]In 1999, some updated contributions were included in a Special Section on Massive Datasets by the *Journal of Computational and Graphical Statistics.* This chapter is a moderately revised version of Huber (1999) and is reprinted with permission from the JCGS. ©1999 by the American Statistical Association. All rights reserved.

3.1   INTRODUCTION

This paper collects some of my observations at, reactions to, and conclusions from the workshop on Massive Data Sets in Washington D.C., July 7-8, 1995.[2] We had not gotten as far as I had hoped. We had discussed long wish-lists, but had not winnowed them down to a list of challenges. While some position papers had discussed specific bottlenecks, or had recounted actual experiences with methods that worked, and things one would have liked to do but couldn't, those examples had not been elaborated upon and inserted into a coherent framework. In particular, the discussions in the Small Groups barely had scratched the implications of the fact that massive sets differ from smaller ones not only by size. Maybe an additional day, providing more time for thinking and for informal contacts and discussions, would have been beneficial. I shall try to continue the discussion of some of the points we left unfinished and connect some of the open ends.

The workshop had been preceded by an intensive electronic exchange of position papers and comments, and this may have helped the talks and discussions during the two days in Washington to stay remarkably free of obfuscating hype. The discussions were more penetrating and looked farther ahead into the future and with more acuity than those at any of the meetings on Massive Data, Data Mining and Knowledge Discovery in Databases I have attended since then. The latter, however, have added mutual understanding, or perhaps more accurately: they have delineated the areas of misunderstanding between statisticians and data base people.

By the late 1990s, Data Mining became the fashion word of the decade and was touted as a cure-all for the problems caused by data glut. In the subsequent decade some of the hype has abated. Most of the so-called data mining tools are nothing more than plain and simple, good old-fashioned methods of statistics, with a fancier terminology and in a glossier wrapping. What made those methods work in the first place, namely the common sense of a good old-fashioned statistician applying them, did not fit into supposedly fully automated, "all artificial, no natural ingredients" packages. Unfortunately, the resulting pattern – a traditional product in a different package, combined with hard sell and exaggerated claims – matches that of snake oil. But in fairness I must add that Usama Fayyad and the work done at JPL persuaded me that there is a much more interesting and promising aspect, also covered by the overall umbrella of Data Mining. It can be characterized as sophisticated pre-processing of massive data sets, in order to scale things down to a size fit for human consumption.

---

[2] Papers appearing in the Proceedings of the workshop (see Kettenring and Pregibon 1996) are referenced by [W].

3.2 DISCLOSURE: PERSONAL EXPERIENCES

Clearly, the personal viewpoints of the workshop participants were heavily influenced by the data sets they had worked with. We somehow resembled the proverbial group of blind men confronted with an elephant. This makes it mandatory to disclose the data that have shaped one's views. In my case these were: electroencephalogram data, children's growth data, census data, air traffic radar data, environmental data, hospital data, marketing research data, road quality data, agricultural and meteorological data, with sizes ranging from 3 Mbytes to 2 Gbytes. Most data sets were observational, a few were opportunistic; there were no imaging data. The census data were an outlier in several respects. I shall later cite specific examples for illustrative purposes. Perhaps the most important thing I have learned from these experiences was: even though the data sources and the analysis goals at first blush seemed disparate, the analysis of a massive data set almost invariably converged toward a sometimes rudimentary, sometimes elaborate, customized data analysis system adapted to the particular set. The reason of course is that in the case of large data sets many people will have to work for an extended period of time with the same or similar data.

3.3 WHAT IS MASSIVE? A CLASSIFICATION OF SIZE

A thing is massive, if it is too heavy to be moved easily. We may call a data set massive, if its mere size causes aggravation. Of course, any such a characterization is subjective and depends on the task, one's skills, and on the available computing resources.

In my position paper (Huber 1994b), I had proposed a crude objective classification of data by size, from *tiny* (10^2 bytes), *small* (10^4), *medium* (10^6), *large* (10^8), *huge* (10^{10}) to *monster* (10^{12}). The step size 100 is large enough to turn quantitative differences into qualitative ones: specific tasks begin to hurt at well defined steps of the ladder. Whether monster sets should be regarded as legitimate objects of data analysis is debatable (at first, I had deliberately omitted the "monster" category, then Ed Wegman (1995) added it under the name "ridiculous"). Ralph Kahn's description of the Earth Observing System however furnishes a good argument in favor of planning for data analysis (rather than mere data processing) of monster sets [W].

Data analysis goes beyond data processing and ranges from data analysis in the strict sense (non-automated, requiring human judgment based on information contained in the data, and therefore done in interactive mode, if feasible) to mere data processing (automated, not requiring such judgment). The boundary line is blurred, most data analytic work includes some data processing (for example simulations done in batch mode), and parts of a judgmental analysis may later be turned into unsupervised preparation of the data for later analysis, that is, into data processing. For example, most of the tasks described by Bill Eddy in connection with fNMR imaging must be classified as data processing [W].

3.4   OBSTACLES TO SCALING

By the mid 1990s we had gained considerable experience with data analysis of small and medium sets. The PCs of that time were excellently matched to the requirements of interactive analysis of medium sets – if one tried to go beyond, one would hit several bottlenecks all at once. Some hard obstacles were caused by human limitations, by computational complexity or by technological limits. Others were financial (hardware costs), or lack of software (e.g., for massive parallelism). Since then, overall computer performance has increased by roughly a factor 100. This may, but need not, imply that we can scale up from medium to large data sets, that is, from 10^6 to 10^8 bytes. My impression is that the increased performance has added convenience rather than power. A brief discussion of obstacles to scaling is indicated – what obstacles we have met and are likely to meet in the immediate future.

3.4.1   Human limitations: visualization

Direct visualization of a whole data set through scatterplots, scatterplot matrices and similar devices is feasible without significant information loss up to about medium sets. This limitation has to do with the resolution of the eye and with the signal processing by the human visual system. It holds across the board, also for imaging data: an efficiently coded high-resolution picture comprises at most a few megabytes. Wegman (1995) elaborates on this. If one wants to go beyond medium size sets, one must either step through subsets or show summaries, with information loss (e.g. density estimates). For large sets (10^8 bytes) and beyond, the human visual system and human endurance put an end to loss-less exhaustive visualization.

In short: Visualization for large data is an oxymoron – the art is to reduce size before one visualizes. The contradiction (and challenge) is that we may need to visualize first in order to find out how to reduce size.

3.4.2 Human – machine interactions

Necessary prerequisites for interactivity (that is, for data analysis in the strict sense, as against mere data processing) are: the task is such that a sequence of reasonably straightforward decisions have to be made in relatively quick succession, each of them based on the results of the preceding step. All three parts of this prerequisite can be violated for large sets: the decisions may be not straightforward because of data complexity, the response may be too slow (the human side of the feedback loop is broken if response time exceeds the order of human think time, with the latter depending on the task under consideration), and it may be difficult to provide a rational basis for the next decision if one cannot visualize the preceding results.

In interactive work, the timing requirements are stringent: For high-interaction graphics the response time must be a fraction of a second, for most other tasks of interactive data analysis it can be a few seconds, but it may exceed 10-20 seconds only very rarely.

In short: Interactivity requires a short, and even more importantly, a predictable response time. This can only be achieved by single-user computers, that is by PCs or workstations, with the latter class now being squeezed out by the former. For all practical purposes such a requirement excludes supercomputers. And if we insist on nearly exhaustive visual inspection, we cannot scale up beyond large sets, or 10^8 bytes.

The following subsections apply only to the data processing aspects of data analysis.

3.4.3 Storage requirements

Backup storage (disk) must be large enough to hold the raw data plus several derived sets. For comfortable work it ought to provide space for the equivalent of at least 10 copies of the raw data set. Nowadays this hardly ever will constitute a critical obstacle.

On a single-processor machine with flat high-speed memory, the latter must be large enough to hold 4 copies of the largest array one intends to work with (otherwise one runs into severe swapping problems with two-argument array operations such as matrix multiplications); for comfortable work, it ought to be at least twice as large. Thus, in order to run the processor at full speed, one may need almost as much free memory as free disk space. *At present [2010], this may become a critical limitation on PCs if we want to scale up operations to huge sets (10^{10} bytes, or 10 gigabytes) and beyond.*

3.4.4 Computational complexity

Processor speed does not scale well, since computational complexity (defined as the number of operations used when the algorithm is applied to a data set of size n) tends to increase faster than linearly with data size.

Fourier transforms and simple standard data base management tasks such as sorting have low computational complexity $O(n \log(n))$ and remain feasible in terms of operations count beyond terabyte monster sets in batch mode even on PCs.

Others (e.g. most clustering algorithms) have computational complexity near $O(n^2)$, so that data sets that are 100 times larger require an increase in computer performance by 10000. Operation counts may become excessive in the large to huge range, even for supercomputers. Note that $O(n^2)$ does not yet correspond to what one customarily calls "computer intensive" algorithms. Simple order-of-magnitude calculations show that computer-intensive operations on the whole of a huge set are infeasible even when the data has a coarse-grained structure and the intensive operations, say being of the order $O(n^3)$, are restricted to one grain at a time, unless those grains are quite small.

The typical operations of linear algebra have complexity near $O(n^{3/2})$, and probably this is what one encounters most in data analysis. If the data is arranged in the form of a matrix with r rows and c columns, $n = rc$, with $r > \sqrt{n} > c$, then tasks with complexity $O(nc)$, such as multiple regression, singular value decomposition and multiplication with a $c \times c$ matrix on the right, all stay within $O(n^{3/2})$, while clustering algorithms, with complexity of at least $O(nr)$, are above.

However, arguments about computational complexity, that is about operation counts, may be eclipsed by those about data flow problems. At present [2010], there are supercomputers with peak performances in the teraflop (10^{12}) to petaflop (10^{15}) range. It is projected that exaflop computers (10^{18} floating point operations, or flops, per second) shall be built by 2019. The hyped-up boasts about peak performance ignore all questions of sustained performance and data flow, whose just-in-time management will become a progressively more vexing bottleneck, considering that light moves a mere 0.3 mm in 10^{-12} seconds. It seems to me that only very special applications, such as fluid dynamics or weather prediction, with very special software, will ever be able to take advantage of such processor performances, while general purpose applications, such as most of those occurring in data analysis, will be bogged down by data flow at the latest in the teraflop range. Since high peak performances usually are achieved with the help of massive parallelism, one runs in addition into severe software problems, see also Section 3.10.2.

3.4.5 Conclusions

Because of human limitations, high interactivity becomes infeasible already for data sizes in the medium (10^6) to large range (10^8 bytes). We then have to switch to the "Lunar Mascons paradigm": form a hypothesis, build a model, do extensive data processing in batch mode, inspect the results (see Section 2.3).

Algorithms with computational complexity above $O(n^{3/2})$ do not scale well. In the large to huge range (10^{10} bytes), tasks with complexity of $O(n^2)$ – which do not yet count as "computer intensive" in today's jargon – become infeasible even on future supercomputers.

3.5 ON THE STRUCTURE OF LARGE DATA SETS

Larger data sets tend to be structurally different from smaller ones. They are not just more of the same, they are larger because they have to be larger. In particular, they are, as a rule, much more heterogeneous.

The Internet has emerged as a fitting paradigm for many of the issues to be discussed here: Internet data are massive, heterogeneous, messy, opportunistic, distributed, and unorganized. The well-known, annoying problems of global internet searches – e.g. finding too many items but not the relevant ones – get magnified into potentially fatal concealed bias problems when one is trying to extract a subset "relevant" for a particular data analysis from a heterogeneous massive data collection.

3.5.1 Types of data

Data can be experimental (from a designed experiment), observational (with little or no control of the process generating the data), or opportunistic (the data have been collected for an unrelated purpose). Massive data sets rarely belong to the first category, since by a clever design the data flow often can be reduced already before it is recorded (of course there are exceptions, e.g. computational fluid dynamics). But they often belong to the third category for plain reasons of economy.

Sometimes, data sets are massive because their collection is mandated by law (e.g. census and certain health data), or because they are collected anyway for other purposes (e.g. financial data). Often, however, they have to be massive because smaller sets will not do, and the predominant reason why they will not do is that the data in question are intrinsically heterogeneous. In particular, there may be many observers and many observed objects, both being located in space and time (e.g. aircraft traffic radar data).

I wonder whether the onslaught of massive data sets will finally force us to acknowledge and heed some studiously ignored, but long-standing admonitions going back to Deming (1940), and reiterated by Tukey (1962), to wit: The statistical profession as a whole is paying much too little attention to the need for dealing with heterogeneous data and with data that arise from conditions not in statistical control (randomness).

Even now, hardly any statistics texts mention the pitfalls one encounters when heterogeneous data are grouped indiscriminately ("Simpson's paradox"). A notable exception is Freedman et al. (1991, p. 16ff.). This is a point where statistics teaching, both to future statisticians and non-statisticians, continues to be sorely amiss.

3.5.2   How do data sets grow?

If we think in terms of a hierarchical data organization, data sets may grow by acquiring

- more hierarchical layers, or
- more branches, or
- bigger leaves,

or all of the above. For example, some data sets are extremely large because each single leaf is an image comprising several megabytes.

It must be stressed that actual data sets very often either do not possess a tree structure, or else several conflicting ones. Instead of "leaf", the more neutral terms "case" or "grain" might therefore be more appropriate.

3.5.3   On data organization

Statistical data bases often have a tree structure imposed on them through sampling or data collection (e.g. census districts – housing blocks – households – persons). But there may be several simultaneous conflicting tree structures (e.g. households and employers). Different priority orderings of categorical variables define different tree structures. For very large sets, a clean tree structure is rather the exception than the rule. In particular, those sets often are composed during the analysis from several, originally unrelated sources (for example health data and environmental data, collected independently for different purposes), that are linked as an afterthought through common external (e.g. geographical) references. In our work with some opportunistic data sets, we found that this kind of associative joining of originally unrelated data sets was one of the most important operations. Moreover, the larger the data set is, the more important are subsetting operations, and also these cut across hierarchies or establish new ones.

No single logical structure fits all purposes. In our experience, the flat format traditional in statistics usually turned out to be most expedient: the data are organized as a family of loosely linked matrices, each row corresponding to a "case", with a fixed number of columns, each column or "variable" being of a homogeneous type.

Sometimes, an even simpler linear organization is preferable: a very long unstructured sequence of items, each item consisting of a single observation together with the circumstances under which it was made (who observed whom, which variable, when, where, and the like). From that basis, the interesting parts are extracted as required and restructured into matrix form.

How such a logical organization should be implemented physically is of course an entirely different question. The problem with massive data is to distribute not only the data, but also ancillary materials and retrieval tools over a hierarchy of storage devices, so that the *ad hoc* retrieval and reorganization tasks to be encountered in the course of a data analysis can be performed efficiently. For example, when should one work with pointers, when with copies?

3.5.4   Derived data sets

It has been said that data analysis is a progress through sequences of derived data sets. We can distinguish between at least four levels of derived data sets:

- raw data set: rarely accessed, never modified,

- base data set: frequently accessed, rarely modified,

- low level derived sets: semi-permanent,

- high level derived sets: transient.

The base set is a cleaned and reorganized version of the raw set, streamlined for fast access and easy handling. The base set and low level derived sets ordinarily must be maintained on some mass storage device for reasons of space. Their preparation may involve sophisticated and complex, time-consuming data processing. The highest level derived sets almost by definition must fit into high speed memory for reasons of computational efficiency. The actual sizes and details of organization clearly will be governed by the available hardware and software. To fix the idea: on a computer doing approximately 100 Mflops and having a few Gbytes of free memory (this corresponds roughly to a PC of 2010), one may just be able to handle a huge raw data set (10 Gbytes). In this case, high level derived sets might comprise about 100 Mbytes of data each for non-graphical tasks, but at most a few Mbytes for tasks involving highly interactive visualization. With massively parallel hardware some of these figures can be pushed higher, but adequate software for data analysis does not exist. More comments on this are contained in Section 3.10.

Derived sets can be formed in various ways. In our experience, low level derived sets mostly are created by application specific preprocessing, or by subsetting (more about this in Sections 3.8 and 3.10). Summaries are problematic with large sets – one ought not to group or summarize across heterogeneity – and splitting into homogeneous parts may be an overly expensive clustering problem. Thus, one will be restricted in practice to splitting based on external *a priori* information or, if data based, to CART-like single-variable methods (for CART see Breiman et al. 1984). Afterwards, summaries of the homogeneous parts then may be recombined into new derived sets.

3.6 DATA BASE MANAGEMENT AND RELATED ISSUES

Data base type operations get both harder and more important with larger sets, and they are used more frequently. With small and medium sets, where everything fits into high speed memory with room to spare and where all tasks are easily handled by a single processor, one does not even realize when one is performing a data base operation on the side. But larger sets may have to be spread over several hierarchical storage levels, each level possibly being split into several branches. Parallel processors and distributed memory create additional complications. With large sets, processing time problems have to do more with storage access than with processor speed. To counteract that, one will have to produce small, possibly distributed, derived sets that selectively contain the required information and can be accessed quickly, rather than to work with pointers to the original, larger sets, even if this increases the total required storage space and creates tricky problems with keeping data integrity (e.g. with carrying back and expanding to a superset some changes one has made in a subset).

In view of the importance and central role of data base operations, it has been suggested that future data analysis (DA) systems should be built around a data base management (DBM) kernel. But paradoxically, all the usual DBM systems do a very poor job with large statistical data bases. For an explanation why this is so, see French (1995), who confronts the design goals of the ordinary DBM systems with those of decision support systems (DSS). Data analysis needs all facilities of a DSS, but more flexibility, in particular read-write symmetry to assist with the creation and manipulation of derived sets. As a consequence, the designer of a DA system must perforce also design and implement his or her own DBM subsystem.

To get a good grip on those problems, we must identify, categorize and rank the tasks we actually perform now with moderately sized sets. We then must identify specific tasks that become harder, or more important, or both, with massive data sets, or with distributed processors and memory. In any case, one will need general, efficient subset operations that can operate on potentially very large base sets sitting on relatively slow storage devices.

Data base management people as a rule do not understand that data analysts/statisticians have requirements rather different from those they themselves are used to. It is worthwhile to spell out these misunderstandings. The DB community paradigmatically operates on one data base (which is constantly updated). The DA community operates on various derived sets, and would like to lock the base set underlying those derived sets during the actual analysis. If the results of two or more passes through the data set have to be compared or combined, a change in the base can create havoc: you will never know whether a shift in the results is due to a change in the analysis procedure or a shift in the base. Assume for example that in the first pass you calculate the mean $\bar{x} = \sum x_i / n$, in the second pass the sample variance $s^2 = \sum (x_i - \bar{x})^2 / (n-1)$, then you are not allowed to change the x_i in between! (I am of course aware that this example is too simple, numerically stable one-pass algorithms are available in this case.)

In sharp contrast to the operations common in data base management, statistical operations typically need to access most pages of the data base anyway. Therefore, processing time will be directly proportional to physical file size. Since most data analyses will extend over several sessions, each session involving many statistical operations and possibly several false starts, it is usually worthwhile to reduce size by preparing an analysis-specific derived set (or sets), to be used during one or more analysis sessions. Collectively, we refer to such semi-permanent derived set(s) as the " base set". Its purpose is similar to that of a "Data Warehouse": it helps to avoid painful dealings with the raw data.

Steps in the preparation of the base set: We usually rewrite the data base into a smaller analysis-specific derived set (i) by eliminating those variables and cases that predictably will not be used in the particular analysis sessions; (ii) by recoding 0-1 variables into single bits; (iii) by aggregating identical records (or similar records after grouping) into a single record plus a count. Typically, this is done not by using the DBM system, but by reading and rewriting the DB files themselves with the help of programs of our own. We would not do it this way if the DBM systems had as flexible export facilities as we by now have had to devise for import! Occasionally, but not always, one will deal with missing values already at this stage. The preparation of derived sets seems to run against the grain of the DB community, but is an essential ingredient in the data analysis process.

For statistical purposes, most data bases are thought of as being logically structured as matrices x(i,j), whose elements are "values" (possibly highly structured by themselves), with i identifying the "case" and j identifying the "variable". Since many operations will access all (or most) cases, but only few variables at a time, it may be worthwhile to transpose the matrix and to store it column-wise for faster access. Moreover, in data analysis the set of "cases" tends to be much more stable than the set of "variables". This is opposite to the situation in DBM. In DA it is very common to create new derived variables, and a column-wise organization makes it easier to append new variables. It must be possible to alternate without undue pain between row- and column-wise representations.

The operations to be performed on the base set are not predictable. They may be complex and may require non-trivial ad hoc *programming. Missing values can be a particular headache and may require specialized tricks (multiple imputation, EM algorithm, survival analysis, etc.).*

3.6.1   Data archiving

There is a dire need for active, rather than merely passive maintenance of data collections. Over the decades, we have witnessed the rise and fall of dozens of storage media and formats: various sorts of paper tapes, punch cards, magnetic tapes, floppies, etc., and I am not alone in having lost a fair number of data collections to media obsolescence because I had neglected to transfer them in time to a currently fashionable carrier. Modern data carriers become technically obsolete so rapidly that nowadays data collections are irretrievably lost after gathering dust for 5-10 years. One remembers with nostalgia the more durable ancient media, such as clay tablets, papyrus or parchment, and even ordinary paper.

3.7   THE STAGES OF A DATA ANALYSIS

Most data analysis is done by non-statisticians, and there is much commonality hidden behind a diversity of languages. Rather than to try to squeeze the analysis into a too narrow view of what statistics is all about, statisticians ought to take advantage of the situation, get involved interdisciplinarily, learn from the experience, expand their own mind, and thereby their field, and act as catalysts for the dissemination of insights and methodologies. Moreover, the larger the data sets are, the more important the general science aspects of the analysis seem to become relative to the "statistical" aspects.

I believe that some of the discussions at the workshop have become derailed precisely because they were too much concerned with categories defined in terms of classical statistical concepts. In retrospect, it seems to me that it might have been more profitable to structure the discussions according to stages common to most data analyses and to watch out for problems that become more pronounced with more massive data.

At the risk of belaboring the obvious, I am providing a kind of commented checklist on steps to be watched.

3.7.1   Planning the data collection

Very often, the data is already there, and one cannot influence its collection and its documentation any more.

The planning of a large scale data collection runs into problems known from Big Science projects: many different people are involved over several years in a kind of relay race. By the time the data are ready to be analyzed, the original designers of the experiment have left or are no longer interested, and the original goals may have been modified beyond recognition.

The obvious conclusion is that big scale data collection must be planned with an open mind for unforeseen modes of use.

Beware of obsolescence. The documentation must be complete and self-sufficient – 10 years later, technical specifications of the measuring equipment may be lost, and names of geographical locations and the scope of ZIP code numbers may have changed. Even the equipment to read the original media may be gone.

If one is planning to collect massive data, one should never forget to reserve a certain percentage of the total budget for data analysis and for data presentation.

3.7.2   Actual collection

It is not possible to plan and specify correctly all details ahead. In particular, minor but crucial changes in the coding of the data often remain undocumented and must afterwards be reconstructed through painstaking detective work. Whoever is responsible for collecting the data must also be held responsible for documenting changes to the code book and keeping it up-to-date.

Everybody seems to be aware of the need for quality control, in particular with regard to measurement instrument drift and the need for continuous calibration. There is much less awareness that also the quality of hardware, software and firmware of the recording system must be closely watched. I have personally encountered at least two unrelated instances where leading bits were lost due to integer overflow, in one case because the subject matter scientist had underestimated the range of a variable, in the other case because a programmer had overlooked that short integers do not suffice to count the seconds in a day. I also remember a case of unusable data summaries calculated on-line by the recording apparatus (we noticed the programming error only because the maximum occasionally fell below the average).

3.7.3   Data access

As data analysts we need tools to read raw data in arbitrary and possibly weird binary formats. An example was given by Allen McIntosh in connection with Telephone Network Data [W]. Not only the actual reading must be efficient, but also the *ad hoc* programming of data input must be straightforward and easy; we have repeatedly run into such problems and have found that very often we were hunting a moving target of changing data formats.

In addition, we must be able to write data in similarly weird formats, in order that we can force heterogeneous sets into a homogeneous form.

3.7.4   Initial data checking

Usually, the problem is viewed as one of legality and plausibility controls. What is outside of the plausible range is turned into missing values by the checking routine.

This is a well-tested, successful recipe for overlooking obvious, unexpected features, such as the ozone hole (which was discovered several years late because "implausibly low" values had been suppressed as errors).

The real problem of data checking has to do with finding systematic errors in the data collection, and this is much harder! For example, how does one find accidental omissions or duplications of entire batches of data? The "linear" data organization mentioned in Section 3.5.3 facilitates such checks. Error checking is never really finished – some insidious error can turn up at any stage.

3.7.5   Data analysis proper

A person analyzing data alternates in no particular sequence between the following types of activities:

- Inspection

- Error checking

- Modification

- Comparison

- Modeling and Model fitting

- Simulation

- What-if analyses

- Interpretation

With massive data sets, both the inspection and the comparison parts run into problems with visualization. Interpretation is thinking in models. Models are the domain of subject matter specialists, not of statisticians; not all models are stochastic! Therefore, modeling is one of the areas least amenable to a unified treatment and thus poses some special challenges with regard to its integration into general purpose data analysis software through export and import of derived sets.

3.7.6   The final product: presentation of arguments and conclusions

With massive data, also the results of an analysis are likely to be massive. Jim Hodges takes the final product of an analysis to be an argument. I like this idea, but regard it a gross oversimplification: in the case of massive data we are dealing not with a single argument, but with a massive plural of arguments. For example with marketing data, a few hundred persons may be interested in specific arguments about their own part of the world, and once they become interested also in comparisons ("How is my product X doing in comparison to product Y of my competitor?"), complexity grows out of hand. However, there is a distinction between potential and actual: from a near infinity of potential arguments, only a much smaller, but unpredictable, selection will ever be actually used.

With massive data, the number of potential arguments is too large for the traditional pre-canned presentation in the form of a report. One rather must prepare a true decision support system, that is a customized, special-purpose data analysis system sitting on top of a suitable derived data set that is able to produce and present those arguments that the end user will need as a basis for his or her conclusions and decisions. If such a system does a significantly better job than, say, a 1000-page report, everybody will be happy; this is a modest goal. See also Section 2.5.9

3.8 EXAMPLES AND SOME THOUGHTS ON STRATEGY

By now, we have ample experience with the analysis of medium size data sets (data in the low megabyte range), and we begin to feel reasonably comfortable with large sets (10^8 bytes, or 100 megabytes), even though direct visualization of larger than medium sets in their entirety is an unsolved (and possibly unsolvable) problem. Let us postulate for the sake of the argument – somewhat optimistically – that we know how to deal with large sets.

Assume you are confronted with a huge data set (10^{10} bytes, or 10 gigabytes). If a meaningful analysis is possible with a 1% random subsample, the problem is solved – we are back to large sets. Except for validation and confirmation, we might not even need the other 99% .

Assume therefore that random samples do not work for the problem under consideration. They may not work for one of several possible reasons: either because the data are very inhomogeneous, or because they are highly structured, or because one is looking for rare events, or any combination of the above. In all these cases, summary statistics – whether in tabular or in graphical form (histograms, density estimates, box plots, ...) – will not work either.

Example: Air traffic radar data. A typical situation is: some 6 radar stations observe several hundred aircraft, producing a 64-byte record per radar per aircraft per antenna turn, approximately 50 megabytes per hour. If one is to investigate a near collision, one extracts a subset, defined by a window in space and time surrounding the critical event. If one is to investigate reliability and accuracy of radars under real-life air traffic conditions, one must differentiate between gross errors and random measurement errors. Outlier detection and interpretation is highly non-trivial to begin with. Essentially, one must first connect thousands of dots to individual flight paths (technically, this amounts to tricky prediction and identification problems). The remaining dots are outliers, which then must be sorted out and identified according to their likely causes (a swarm of birds, a misrecorded azimuth measurement, etc. etc.). In order to assess the measurement accuracy, one must compare individual measurements of single radars to flight paths determined from all radars, interpolated for that particular moment of time. Summary statistics do not enter at all, except at the very end, when the results are summarized for presentation.

Exhibit 3.1 may give an impression of the complexity and messiness of the situation. It shows a small subset (well below 1%) of a one hour collection of radar data, namely the radar tracks of 40 airplanes, observed by one radar station. The radar station is represented in the center of the picture by a little icon. The original range-azimuth data were transformed into cartesian coordinates. Much of the available data is not shown, in particular time and altitude information, the (incomplete) identification of the planes by transponder data, and ancillary data, such as a geographical map with airport locations. The tracks near the maximum range of the station tend to get messy and exhibit both gaps and outliers. Other irregularities are caused by topographic features or electronic interference.

I believe this example is typical: the analysis of large sets either begins with task and subject matter specific, complex preprocessing, or by extracting systematic subsets on the basis of *a priori* considerations, or a combination of the two. Summaries have no place at the beginning, they enter only later, in connection with interpretation and presentation of results. Often, the subsets will be defined by windows in space and time. Even more often, the selection has two stages: locate remarkable features by searching for exceptional values of certain variables, then extract all data in the immediate neighborhoods of such features. For a non-trivial example of preprocessing, compare Ralph Kahn's description of the Earth Observing System [W] and the construction of several layers of derived data sets. For one beginning with subsetting, see Eric Lander's description of how a geneticist will find the genes responsible for a particular disease: in a first step, the location in the human genome (which is a huge data set, 3×10^9 base pairs) is narrowed down by a factor 1000 by a technique called genetic mapping (Lander 1995).

X

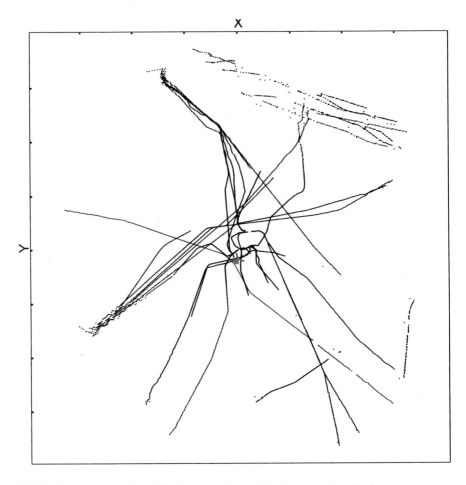

Exhibit 3.1   Radar tracks of 40 airplanes, observed for about one hour by the grey radar station in the center. Each dot corresponds to one observation of one plane. The tickmarks are spaced 50 km.

After the preparatory steps, one may want to look up additional information in other data bases, possibly from informal external sources:

> **Example: Environmental data.** We found (through exploratory data analysis of a large environmental data set) that very high radon levels were tightly localized and occurred in houses sitting on the locations of old mine shafts.

In this example, indiscriminate grouping would have hidden the problem and would have made it impossible to investigate causes and necessary remedies. The issue here is one of "data mining" in the sense of looking for a rare nugget, not one of looking, like a traditional statistician, "for a central tendency, a measure of variability, measures of pairwise association between a number of variables". Random samples would have been useless, too: either one would have missed the exceptional values altogether, or one would have thrown them out as outliers.

Data analysis is detective work. The metaphor is trite but accurate. A careful distinction between tasks requiring the acumen of a first rate sleuth and tasks involving mere routine work is required. After perusing some of the literature on data mining, I have begun to wonder: too much emphasis is put on futile attempts to automate non-routine tasks, and not enough effort is spent on facilitating routine work.

In particular, everybody would like to identify noteworthy, but otherwise unspecified features by machine. From my experience with projection pursuit on small to medium sets I think this is a hopeless search for the holy Grail (computational complexity grows too fast with dimensionality). Pattern discovery is intrinsically harder than pattern recognition. A less ambitious, still hard task is the approximate match-up problem: find all structures in data set A that are approximately similar to some structure in data set B, where A and B are large. It is not at all clear whether even such problems can be solved within the desired $O(n^{3/2})$-complexity.

See Chapter 2 for further thoughts on strategy.

3.9 VOLUME REDUCTION

Volume reduction through data compression (with information loss) sometimes is advocated as a kind of panacea against data glut: "keep only what is exceptional, and summarize the rest". At least in the case of observational, as against experimental data, I think this is a daydream, possibly running counter to several of the reasons why the data are being collected in the first place! It is only a slight exaggeration to claim that observational data deserve to be saved either completely or else not at all. For example, a survey of the sky must be preserved completely if one later wants to check the early stages of supernovae.

But prior to *specific* analyses, targeted volume reduction usually can be performed on a data matrix either by reducing the number of rows (cases) or the number of columns (variables), or both. This might be done based on *a priori* interest. Data-driven reduction might be done for example by aggregation (i.e. by combining similar rows) or by summarizing (e.g. by forming means and variances over a homogeneous set of rows). Reducing the number of variables is usually called "dimension reduction" and can be done for example by variable selection (pick one of several highly correlated columns), or by forming (linear or nonlinear) combinations of variables.

But it is difficult to put the general notion of dimension reduction on a sound theoretical basis; exploratory projection pursuit comes closest, but as its computational complexity increases exponentially with dimension, it is not well suited to massive data sets.

The next best general approach is dimension reduction through principal component or correspondence analysis (i.e. the truncated singular value decomposition): remember that the k leading singular values yield the best approximation (in the square norm sense) to the data matrix by a matrix of rank k. According to Sue Dumais this was surprisingly successful in the context of information retrieval even with quite large data matrices [W].

However, the most powerful type of dimension reduction is through fitting local models:

> **Example: Children's growth data.** Several hundred children were observed periodically from birth to adulthood. Among other things, for each child, 36 measurements of body length were available. It was possible to reduce dimension from 36 to 6, by fitting a 6-parameter growth curve to each child. The functional form of that curve had several components and had been found by estimating some 30 global parameters from the total available population of children. Most of the 6 parameters specific to a particular child had an intuitive interpretation (age at puberty, duration and intensity of pubertal growth spurt, etc.); the residual error of the fit was only slightly larger than the intrinsic variability of the measurements. (Stuetzle et al. 1980.) *See Section 8.2.*

Local model fitting is computationally expensive, but typically, it seems to stay within the critical $O(n^{3/2})$-limit.

3.10   SUPERCOMPUTERS AND SOFTWARE CHALLENGES

On the super-workstations available in the mid 1990s, or on the PCs available in 2010 (100 Mflops, 1 Gbyte of free memory) one can certainly handle large sets with almost-present-day software, and one may just barely be able to handle huge sets (10 Gbytes). To push those figures substantially higher, one would have to invest in massively parallel supercomputers and novel software. Is such an investment worthwhile? What are its chances to succeed? I do not have final answers, but would like to offer some food for thought.

3.10.1   When do we need a Concorde?

I believe the perceived (or claimed) need for supercomputers exceeds the real need.

The problem is that of the defunct Concorde: flying the Concorde is a status symbol, but if too much time is spent on the ground, the fast flight is not worth the money. Response times in fractions of a second are neither needed nor appreciated, if it takes several minutes, and possibly hours, to think up a question and to digest the answer.

We must learn to identify and differentiate between situations where supercomputers are necessary or at least truly advantageous, and situations where they are not.

In the 1990s we had encountered several examples with raw data sets in the 10 to 900 MByte range, where it had been claimed that a large mainframe or supercomputer plus several months of customized programming were needed in order to accomplish certain data analytic tasks. In all those cases we found that a well endowed PC, plus a few weeks of customized programming in a high-level interpreted data analysis language (ISP) would be adequate, with comparable or even shorter execution times, at total costs that were orders of magnitude lower.

3.10.2   General Purpose Data Analysis and Supercomputers

Can general purpose data analysis take advantage of supercomputers? In view of some remarks made in Section 3.4, this applies only to the data processing aspects of data analysis. Dongarra's famous series of benchmark comparisons (when I wrote this, I had the version of April 13, 1995 in front of me) highlights the crux of the problem: without special hand-tuning efforts to improve and distribute the data flow, the fastest multi-processor supercomputers beat the fastest single-processor super-workstations merely by a factor 4, which is not worth the money. Tuning may yield another factor 20 or so. We need to discuss strategies for recovering that factor 20. *Ad hoc* code tweaking on a case by case basis is so labor intensive and error-prone that ordinarily it will be out of the question. That is, we have a very serious software challenge. *Nowadays [2010] this software challenge arises already on the PC level – high end PCs try to boost performance by increasing the number of processing units.*

Our experiences with ISP suggest how a speed-up could be done even in a general purpose system that is geared toward improvising ever new applications. ISP is a small, general purpose, array oriented, interpretive data analysis language somewhat similar to S. It contains a core of 100 to 200 building blocks (commands or functions).

Explicit, interpreted looping is slow, but it is rarely needed. Efficiency relative to compiled code increases with data size because of array orientedness. I believe that many of the building blocks can be beefed up for parallelism, but there may be snags. For example, reorganization and redistribution of data between subtasks might be more expensive than anticipated.

The proof of the pudding is in the eating, i.e. we need to conduct the experiment suggested by Huber (1994b, at the end of section 8) and build a working pilot system. To prove the point we should aim for a small, relatively unspecific, but universal system. For a basis, I would choose ISP over S precisely because it has these three properties. But we should take the opportunity to build a new, better system from scratch, rather than trying to port an existing system. From past experiences I estimate that it will take three years before the pilot system for such an experiment attains sufficient maturity for beta testing. We better begin soon.

Regrettably, this never happened. But in retrospect I wonder whether I had overrated the multi-processor software issue, at least with regard to common data analytic applications. Perhaps the small-grained parallelism I had in mind when I wrote the above is not that much relevant. In my experience, the most time-consuming data analytic applications occurred in connection with tasks such as simulation, where it might be cheaper to farm out the task to multiple processing units, each having its own copy of the program and its own large private chunk of memory.

3.10.3   Languages, Programming Environments and Data-based Prototyping

Here is an account of some things we have learned from our ISP experience (see also Chapter 4). The traditional programming languages, (FORTRAN, C, PASCAL, ...) are too low-level for the purposes of data analysis. We learned that already back in the 1970's: there is much re-programming and re-use under just slightly different circumstances, and for that, these languages are too clumsy and too error-prone (cf. also the comments by Allen McIntosh [W]). Subroutine libraries like LINPACK help, but are not enough. We need a **high-level array oriented language** on top, with a simple syntax and safe semantics, whose units or building blocks must be very carefully selected for universal use. The language must be user-extensible through combination of those units into new building blocks with the same syntax. The core building blocks must be highly optimized.

In the 1980's we became aware of the need for **programming environments** in the style of SMALLTALK and LISP machines. In data analysis, you never hit it

)

right the first, or the second, or even the third time around, and it must be possible to play interactively with modifications, but without having to start everything from scratch. Rather than building a system on top of SMALLTALK or LISP, we decided to augment our data analysis language ISP so that it acquired its own programming environment.

Around 1990, we realized that we had to go even further into what we call **data-based prototyping**: build a customized data analysis system while actually doing production work with the data. The basic problem is that the user (whether it is a customer or we ourselves) never is able to specify the requirements in advance. Our solution is to mock up a rough-and-ready working prototype, and let the user work with it on his or her actual data. Without involving the actual user early and actively in the use and re-design of the system, in particular in issues of presentation of results (what to show and how), it is extremely hard to arrive at a satisfactory solution. Some comments made by Schmitz in the context of marketing databases nicely illustrate the difficulty [W]. A high-level language and a good programming environment are indispensable prerequisites for data-based prototyping. See Huber and Nagel (1996).

None of the existing languages and systems is entirely satisfactory. After seeing Carr's list of preferences in Section 4 of his position paper, it seems to me that our own software (ISP) comes closer to an ideal, universal system than I ever would have suspected. It does a job superior to Matlab in the area of visualization. We use it instead of a GIS (geographical information system) because the latter systems have difficulties with discontinuous values that are attached to arbitrary points rather than to grid points or political districts. After becoming dissatisfied with all available general data base software, we began to improvise our own approaches in ISP, and we prefer it to SAS and S for data analysis, especially with large sets. Carr's comment on the "diminished influence of the statistical community upon my work" is reflected by the fact that in ISP we never felt compelled to go beyond a frugal minimum of statistical functions.

3.11 SUMMARY OF CONCLUSIONS

- With the analysis of massive data sets, one has to expect extensive, application- and task-specific preprocessing. We need tools for efficient *ad hoc* programming.

- It is necessary to provide a high-level data analysis language, a programming environment and facilities for data-based prototyping.

- Subset manipulation and other data base operations, in particular the linking of originally unrelated data sets, are very important. We need a data base management system with characteristics rather different from those of a traditional DBMS.

- The need for summaries arises not at the beginning, but toward the end of the analysis.

- Individual massive data sets require customized data analysis systems tailored specifically toward them, first for the analysis, and then for the presentation of results.

- Pay attention to heterogeneity in the data.

- Pay attention to computational complexity; keep it below $O(n^{3/2})$, or forget about the algorithm.

- The main software challenge: we should build a pilot data analysis system working according to the above principles on highly parallel machines.

CHAPTER 4

LANGUAGES FOR DATA ANALYSIS

Computing languages for data analysis and statistics must be able to cover the entire spectrum from improvisation and fast prototyping to the implementation of streamlined, specialized systems for routine analyses. Such languages must not only be interactive but also programmable, and the distinctions between language, operating system and user interface get blurred. The issues are discussed in the context of natural and computer languages, and of the different types of user interfaces (menu, command language, batch). It is argued that while such languages must have a completely general computing language kernel, they will contain surprisingly few items specific to data analysis – the latter items more properly belong to the 'literature' (i.e. the programs) written in the language.[1]

---

[1] This chapter is a moderately revised version of Huber (2000a), reprinted with permission from the *Journal of Computational and Graphical Statistics*. ©2000 by the American Statistical Association. All rights reserved.

4.1   GOALS AND PURPOSES

Over the past 30 years or so, data analysis has emerged as the single most demanding application of interactive computing. The main challenge is that it covers a wide spectrum ranging from research to repetitive routine. Standard tasks should be offered in canned form, in particular to novice users. Non-standard ones should be easy to improvise, either by putting them together from standard building blocks, or by modifying suitable templates, in the style pioneered by the Postscript Cookbook (1985). It should be easy to streamline such improvised tasks and to add them to the system, either in interpreted soft or in compiled hard form (extensibility). After one has done an interactive analysis once, it should be easy to repeat it with slight modifications and with similar data (what-if analyses, and stepping from interactive execution to batch, e.g. for simulation). A user ought to be able to move across a wide spectrum of tasks without hitting steep ramps (i.e. language barriers) in the learning curve.

In short: an ideal system should cover the full range smoothly, from launching canned applications, to improvising new applications, and to canning them. This goal is lofty, and compromises with regard to performance and elegance are unavoidable. But even a modest approximation to this goal will still need the full range of user interfaces, from mouse clicks to command lines and batch programming. While in principle it is possible to design such a system in many different ways, we have argued elsewhere (Huber 1986a, Huber and Huber-Buser 1988) that the most straightforward, and perhaps the only feasible solution is a system built around the middle ground, namely around a programmable, interpreted command language, because from there one can branch out to either side. In the meantime the arguments have gained rather than lost strength, despite the still more fashionable, but overly rigid approach from the extremes: the mouse-click front end and the compiled back end.

Any comprehensive interactive system is crucially dependent on a well-designed interface between humans and machines, mediating a two-way interchange of information between all parties concerned. In a general purpose system, such interfaces will acquire properties customarily associated more with natural languages than with computing systems.

Human-to-machine communication clearly remains the most prominent direction of information exchange, but the three other possibilities are important too: machine-to-machine, machine-to-human and human-to-human.

Machine-to-machine communication occurs for example in connection with graphics hardcopy: Postscript files are written or read by humans only in exceptional cases.

Machine-to-human communication is neglected by many systems. The response by operating systems or compilers to human errors often is pathetic. Some screen designs actually appear to *minimize* the data-to-ink ratio, compare Tufte (1983), who argued that in data graphics one should aim to maximize it, within reason.

Human-to-human communication also is unduly neglected. A good computing language ought to be structured such that it can serve this purpose. Remember that the primary purpose of ALGOL 57 (later to be redesigned into ALGOL 60) had been the complete and unambiguous description of numerical algorithms (whence the name – ALGOrithmic Language). According to oral comments by Rutishauser (who had been the driving force behind the ALGOL effort), compilation by machine was only a secondary consideration: it would prove that the primary goal had been achieved. This attitude is all the more remarkable, considering that Rutishauser's *Habilitationsschrift* (1952) had been the first paper ever written on compiling and compilers, and that Backus, the designer of the original FORmula TRANslation, or FORTRAN (1953), was another member of the ALGOL working group.

The issues are tightly interwoven, and the tale to be told is tangled. The discussion will be arranged in several circles, beginning with a general discussion of languages, both computing and natural. The next circles are concerned with specific interface issues, and finally with the requirements a language suitable for statistics ought to satisfy.

4.2 NATURAL LANGUAGES AND COMPUTING LANGUAGES

> *Diverse lingue, orribili favelle ...*
> (*Diverse languages, horrible utterances ...*
> Dante, Divina Commedia, Inf. 3,25)

> *What can we learn about programming languages from the so-cial aspects of mathematics and natural languages? - Now why is this a relevant topic to consider? Well I definitely feel that there is something rotten in the realm of programming. There is a lot of discussion, but somehow I think that most of it misses the point. There are too many fads, too many quick solutions, a too wide gap between theory and practice.* (P. Naur, 1975)

Sadly, Naur's comments still hold true. Very little attention has been paid by systems developers to what we actually know about everyday natural languages. It is less a question of borrowing directly from natural languages, but rather of borrowing methodology for looking at languages. For example, one ought to pay attention to the syntactic typology of computer languages in the same way one has investigated that of natural languages (compare Winfred P. Lehmann 1978).

4.2.1 Natural languages

The recorded history of natural languages now spans approximately 5000 years. It is difficult to reconstruct precise reasons behind particular developments. But the recorded time span is long enough for natural selection to operate effectively. Here are a few random observations of potential relevance for the design and use of computer languages.

Language skills require extensive training. People stay with the language(s) they have learned, whenever this is possible, and they usually feel most comfortable with their mother's tongue.

Natural languages are mutually translatable into each other. Despite this (or because of this?) difference of language sometimes is the source of deep hate.

Spoken and written forms of a language are consistently being regarded as one and the same language. There are definite differences of style, but a literate person always was supposed to be able to read a text aloud and to take dictation.

Indentation, spaces, capitalization and punctuation serve as visual cues to indicate sentence structure in written texts (they are substitutes for absent phonetic cues). In natural languages, these items carry little, if any, information. The early alphabetic scripts did not use them at all.

Unpronounceable words are hard to memorize. The ancient Egyptian language is difficult to learn not because it is written in hieroglyphs, but because the vowels are not written and therefore unknown.

Human communication is two-way and depends on feedback. We use guesswork and common sense, and if necessary, we ask for clarification or confirmation.

Most alphabetically written languages have a redundancy factor of about two, that is, a text can be reconstructed if up to about half of the letters have been erased (Shannon 1949). If the redundancy deviates too much in either direction, comprehension suffers.

Natural languages avoid deep nesting, except if it is of a trivial chain-type.

The first known writing system was invented 5000 years ago in the Near East. It used pictorial signs ('icons'). Within a few decades, it developed into a full-scale, abstract writing system with hundreds of conventional signs for words (no longer recognizable as pictures of objects) and with phonetic complements to express grammatical relations. About 2000 years later, alphabetic writing systems were invented, allegedly because they were easier for the occasional users who had difficulties remembering several hundred signs.

Icons are successful means of non-verbal communication in a multilingual society, but only if they are small in number, easily recognizable and completely standardized; international traffic signs may be the best example.

4.2.2 Batch languages

Computer science is young and has little sense of history – even of its own history. The time horizon of CS graduates typically does not extend beyond their first programming course. But we can learn some lessons even from the short history of computing languages.

Serious computing began with numerical applications, programmed in raw machine language, and executed in batch mode. I still remember that time (the mid-1950's); in essence, we then wrote programs by informally putting two levels of better readable languages on top of raw machine language: flowchart language and assembly language. We structured the program through flowcharts and then wrote it down by using symbolic names and addresses, instead of actual storage locations. Finally, this program was assembled into machine code by hand. The really painful part was not the *writing* of a program, but *making modifications* to it. The problems with raw machine language were that it was more difficult to read than to write, and that adding a few lines of code usually implied making tricky non-local modifications.

FORTRAN was an impressive first in 1953, but remained an awkward language for almost 25 years, up to the emergence of the FORTRAN 77 standard.

The design of ALGOL 60 and its formal description ('Backus-Naur notation', Naur 1960) was even more impressive; I believe it was Hoare who once said that ALGOL 60 was a notable improvement not only over its predecessors, but also its successors.

Though, ALGOL 60 had a gaping hole: the design committee had not been able to agree on string handling and hence on I/O. This was irrelevant with regard to the primary goal of ALGOL (unambiguous description of algorithms), but not for the secondary (compilation). Commercial ALGOL compilers for a long time were lacking, or poor and inefficient. An overly ambitious and overly complex redesign (ALGOL 68) effectively killed the ALGOL effort.

PASCAL (Wirth 1971) was not a better language, but it was more successful because of the availability of portable compilers. Its problem was that it had been conceived too narrowly, the primary design goal was to teach programming through writing small, well structured programs. Now it is choking under its too narrow conception and its overly strong typing. The problems were evident from the beginning; after looking at its specifications in the early 70's I certainly would not have believed that it would survive into the 90's.

C was developed for the PDP-11 on the UNIX system in 1972. Its strengths are neatly summarized by Ritchie et al. (UNIX 1978, p. 1991): "C is not a 'very high-level' language nor a big one and is not specialized to any particular area of application. Its generality and an absence of restrictions make it more convenient and effective for many tasks than supposedly more powerful languages. [...] The language is sufficiently expressive and efficient to have completely displaced assembly language programming on UNIX."

The weakness of C lies in its crummy syntax, in its error-prone type declarations, and in its unnecessarily risky handling of pointers. As a consequence, it is hard to read and quite unsuited for the communication of algorithms between humans. If I am to venture a prophecy, these problems are serious enough that they will ultimately lead to its downfall and replacement. But this will not happen too soon, compare Naur's (1975) comment that "even a language like Esperanto, that was recognized already in 1907 when it was but 20 years old to be poorly suited for its declared purpose, has retained its support to this day, 67 years later".

It is very difficult for new programming languages to gain acceptance. While special purpose languages continue to blossom, and the better among them (from spreadsheets to Java) continue to be abused for purposes they were never intended for, the design of new general purpose languages has been out of fashion for a while now. Reluctance to learn yet another language is only part of the reason, it also has to do with the fact that language designers never seem to learn from the mistakes corrected by their predecessors. Thus, the principal weakness of FORTRAN was and is its lack of dynamic arrays. ALGOL corrected this in a safe way, but PASCAL fell behind FORTRAN: in standard PASCAL one cannot even write a routine for transposing arbitrary arrays. PASCAL was a very 'safe' language, but C fell even behind FORTRAN in this respect.

The use of computing languages for the unambiguous description of algorithms, to be read also by humans, as with the original ALGOL, has fallen by the wayside. A notable exception is Donald Knuth (see his monumental works on the Art of Computer Programming (1968-) and on Computers and Typesetting (1986)).

4.2.3 Immediate languages

An interactive language is not a language suitable for writing interactive programs, or for writing batch programs interactively, but a language suitable for interactive execution: when the user enters a unit of thought (say a command line), it is immediately executed. Unfortunately, the term 'interactive' is much over-used and misunderstood, so I shall prefer to talk about 'immediate' languages. Among the two best known languages of this class, APL goes back to Iverson (1962), and BASIC arrived only two years later in 1964. What is needed for data analysis is not merely an interactive system, but a *programmable immediate language*.

None of the classical programming languages are suitable for the spectrum of tasks encountered in data analysis. The batch languages (FORTRAN, PASCAL, C, . . .) do not permit immediate use. Those languages that do (APL, BASIC, and shells of various operating systems) fall short in other respects.

APL is write-only: it is easy to write, but it cannot be read, only deciphered. BASIC is not array oriented, and as a consequence, even very simple loops necessitate the writing of a small throw-away program; with many implementations of BASIC (in particular the standard versions of BASIC on the PC), this causes the loss of all previous results obtained in immediate mode.

Among the languages specifically designed for doing statistics and data analysis, S primarily is a high-level programming language. While it permits immediate use, its syntax is poorly suited for it. ISP was designed for immediate use; it was soon extended to include high-level programming features. I believe that among the two it comes closer to the ideal.

Though, talking about Platonic ideals may be out of place here: in order that a language be suitable for data analysis it must compromise its ideals by making concessions. It is more important that it does everything moderately well, than that it does something very well.

With immediate languages the customary distinctions between operating systems, programming languages and user interfaces disappear. There have been several computers whose operating system coincided with their principal programming language, from the tiny Timex-Sinclair using BASIC, to the Symbolics LISP Machine. High-level immediate languages get quite close to the spirit of natural languages in more than one respect, and we will have to single out the similarities and the differences not only to batch programming but also to natural languages.

4.2.4 Language and literature

Naur pointed out already in 1975 (with little success) that there is a need to shift attention away from programming languages toward the corresponding literature, i.e. toward the programs written in the languages. Where does 'language' end and 'literature' begin? In an extensible language, where the extensions look exactly like original language constructs, the boundary is blurred and moving. A possible criterion is: if a particular construct is frequently used as a building block within other constructs, then it belongs to the language. If it is never used in this way, then it belongs to the 'literature' written in the language. If we accept this criterion, then, somewhat surprisingly, the core of a language for data analysis is completely general and contains negligibly few items *specific* to data analysis.

From experiences with our own immediate data analysis language (ISP), a command for multiple linear regression clearly belongs to the language. But apart from minor embellishments, regression is not specific to data analysis: remember that APL has a special, built-in operator doing linear least squares fits. A procedure doing nonlinear least squares is on the border line. A large ISP procedure doing generalized linear models, all of GLIM and a little more, clearly belongs to the 'literature'.

4.2.5 Object orientation and related structural issues

> *'Ergativity' is currently an 'in' term in linguistics. It is used by a wide variety of linguists, with a whole range of different meanings. As a result, much confusion exists at present about what an 'ergative' language is, and about the morphological, syntactic, and semantic consequences of such a characterization.* (R. M. W. Dixon, 1979)

The current shibboleth in programming is OOPLA – Object Oriented Programming LAnguages. It is customarily described by invoking a number of magic words: objects, classes, inheritance, abstraction, encapsulation, polymorphism, and dynamic binding. Apart from that, there is much confusion: the above quote remains accurate if we replace 'ergative' by 'object oriented'.

The nugget hidden in the quote is that ergative natural languages *are* object oriented. They put the sentence focus on the direct object rather than on the subject of the sentence, as all the usual 'accusative' Western languages do. The tell-tale common pattern is that they treat the object of a transitive verb and the subject of an intransitive verb exactly alike, and different from the subject of a transitive sentence. Ergative languages may be in a minority, but completely unrelated ones are known from every corner of the globe; among the better known examples are Eskimo, Georgian, Basque, and among ancient Near Eastern languages, Sumerian and Hurrian. To us native speakers of 'accusative' languages, their syntax is very confusing at first. Some languages have switched in historical times and have become ergative, or have lost ergativity. Curiously, no language is 100% ergative, while there are fully accusative languages. Ergative languages seem to have more than their fair share of grammatical exceptions and irregularities.

The analogy between ergativity and object orientedness of course is not total. In computing languages, the shift of emphasis is not from subject to object, but from verb to object. Nevertheless, there are suggestive similarities. In particular, I wonder whether 100% object orientedness also leads to impossible languages.

Just like ergativity, object orientedness is a matter of degree. By acquiring some more object orientedness, C has turned into C++. SMALLTALK may be the most object oriented language ever created.

Object oriented languages enforce a uniform treatment of related things through inheritance rules. They also may make things simpler for the programmer/user by hiding complexity in a hierarchy of composite structures. The problem is that the complexity is still there, and it occasionally acts up.

Thus, a traditional SMALLTALK example is to print an object by sending it the message 'print yourself'. If the object happens to be composite, what does one expect to be printed:

- a list of the sub-objects?

- all sub-objects?

- the data sub-object, using the format sub-object for formatting?

All three possibilities make perfect sense, and there are some others. In essence, one needs to have at least three distinct variants of the print message, and at any time, one might have to add new ones. For the user, it may be conceptually simpler to spell out things and tell the system exactly what he or she wants to be printed and how.

If one is improvising a data analysis in immediate mode, object orientedness does not help much. On the contrary, it adds one more level that one must pay attention to and that must be debugged: the correctness of an interactively created hierarchy of structures. In immediate languages, too much object orientedness may be just as harmful as too many goto's in batch programming.

The question is: how much is beneficial? Clearly, in an array oriented language the data objects ought at least to carry their own types and dimensions with them. With graphics, data structures get too complex to be handled with ease on a command line: in addition to the data points, there may optionally be lines, areas, colors, symbols, subsets, point labels, variable labels, subset labels, etc. It must be possible to handle the whole as a single entity, but also to operate with ease on the parts, or on groups of the parts. The following anecdote illustrates some of the problems. In ISP, we prototyped such structures as macro procedures, and this approach turned out to be extremely convenient and easy to use. We therefore felt that a streamlined, more sophisticated and more powerful approach in compiled code would be even better. Interestingly, this was not so, the more ambitious approach clashed with the KISS principle ('Keep It Simple, Stupid'): it was not simple and stupid enough to be used in immediate mode! In the macro version, the component objects had been linked simply through having a common family name (e.g. 'quake') followed by an underscore and a two-letter extension (quake_da, quake_li, . . . for data, lines, etc.). There was no actual object 'quake'. In the ambitious version, there was one, sub-objects could be composite too, and the user could interactively create arbitrarily complex structures. However, when even I myself developed problems talking my own language fluently in immediate mode (batch programming was no problem), I stopped the effort during alpha-testing. The probable cause of the problem is that one has to keep an overly large fraction of one's attention focused away from the data and its analysis, namely on remembering *or looking up* structures and on making distinctions between objects and sub-objects in a way one is psychologically not prepared to do.

Moreover, the complexity of an object oriented system increases with the product of the number of command verbs (more accurately: the available types of action) times the number of object classes. If both actions and object classes are user extensible, we have a serious complexity problem, overtaxing the user's memory. It should be emphasized that this is a problem peculiarly acute with immediate languages. Batch environments can tolerate structures that are an order of magnitude more complicated, but I suspect that all the current window systems are exceeding the limit.

4.2.6   Extremism and compromises, slogans and reality

> *Contradictory things [...] are characterised by identity, and consequently can coexist in an entity and transform themselves each into its opposite.* (Mao Tse-tung, 1937)

A language suitable for data analysis must make compromises. Unfortunately, reasonable, but working, compromises never are 'in'. An error in one direction is usually 'corrected' by erring even more into the opposite direction. Here are just a few examples.

Some of the maxims of early UNIX were to 'build afresh rather than complicate', and 'small is beautiful' (UNIX 1978, p. 1902f.). These maxims may be too extreme to be obeyed rigorously: once the installed base reaches a certain size, building afresh becomes difficult. Unfortunately, more recent systems are based not on relaxed versions of those maxims, but on the opposite extremes. For example, a 'Definitive Guide to the X Window System' (X Window 1993) comprises 11 volumes of close to 1000 pages each. In our terminology all this is 'language', not yet 'literature'. Homer, the Bible and Shakespeare together comprise about half as many pages . . .

The absurdity of such developments (and the difficulty to prevent them) was castigated early on by Hoare, even before X Windows were born. In his 1980 Turing Lecture he tells the story of the Emperor who, to avoid nakedness, ordered that each of his many new suits of clothes should simply be draped on top of the old, and of the consulting tailor, who was never able to convince his clients of his dawning realization that their clothes had no Emperor (he had slipped away to become swineherd in another story).

UNIX uses cryptic two-letter command names. The systems built on top indulge in equally unreadable long names such as XmuCopyISOLatin1Uppered.

One of the biggest advances of UNIX (listed as number one in Ritchie's retrospective, UNIX 1978, p. 1947) was to get rid of a multiplicity of file structures. In UNIX, every file is regarded as a featureless, randomly addressable sequence of bytes. The systems now being built on top of UNIX (e.g. the already mentioned X-Windows, plus X-Tools, plus OSF Motif) grow veritable plantations of trees of classes of structures. When there are too many structures, the structure (if there still is any) is obscured.

The original gospel of Structured Programming was to adhere to a strict top-down design: plan the big structure of the application, then fill in the details. This did not work (it is humanly impossible to specify a complex system correctly in advance). The current gospel seems to be the bottom-up approach: design the menu interface first, then fill in the application. Also this approach does not work for the same reasons, and it has led to Potemkin villages: nice looking surfaces whose hidden flaws caused the applications to be still-born.

A rampant type of contradiction has to do with self(?)-delusion: use a name that is opposite to the facts. For example, after AT&T had introduced a so-called 'Portable UNIX', in an actual porting experiment 40% out of a total of 50000 lines had to be changed, namely 16000 lines of C and 4000 lines of assembly code (Jalics and Heines, 1983).

'User-friendliness' is a particularly insidious slogan. It is being used to split the computing community into a caste of priests (the programmers) who have access to the inner sanctum, and a caste of supposedly illiterate users clicking on icons. After the liberating period of the 1980's, when the people gained hands-on control of their own hardware, this is a reversion to the 1960's and 1970's, when the computer centers dictated what was good for the users.

Moral: Watch out. Slogans usually clash with reality.

4.2.7 Some conclusions

From the random remarks made in the preceding sections one can distill a few maxims or rules for language design that should not be broken without good reason:

(1) Aim for a general purpose language.

(2) Keep close to the tested natural language style-sheet.

(3) The language should not only be writable, but also readable.

(4) Use pronounceable keywords.

(5) Do not encode information into case and punctuation.

(6) Avoid icons that are neither standardized nor self-explanatory.

(7) Do not leave holes in a language design (like the I/O in ALGOL 60).

(8) Make the language portable by providing good and portable interpreters or compilers, and by avoiding dependence on hardware and operating systems.

(9) Beware of ambitious redesigns.

4.3 INTERFACE ISSUES

> *But even if the effort of building unnecessarily large systems
> and the cost of memory to contain their code could be ignored,
> the real cost is hidden in the unseen efforts of the innumerable
> programmers trying desperately to understand them and use
> them effectively.* (N. Wirth, 1985)

Languages are for communication, and the primary purpose of a computing language is to provide efficient and effective communication between humans and machines, not to build a monument to one's ingenuity.

When we are talking about a language used to transmit human wishes to the machine, we mean exactly that, and not a machine-internal way of storing those wishes. This point often is misunderstood. Some designers seem to be proud that in their system a complex statement such as

```
MyPanel:Panel  →  InsertObject(
        CopyObject([(3,4)  →  GetValue]))
```

can be generated with a few mouseclicks (Johnson *et al.*, 1993). Actually, it is a reason for worry, it only means that the statement being generated is in a language unsuitable for the human side of man-machine communication. In other words, one just is adding another suit of clothes to cover up an ugly earlier suit.

Still another worry: Voice interfaces are now finally coming of age, thanks to Dragon Systems. Just try to dictate the above statement into the computer. . .

Computing languages can be classified according to the size of the chunks used for transmitting information to the computer. Usually, this is considered an interface issue, but really it is one of language, implying a distinct type of interface:

Small chunks (1-4 bits): menu interface,
Medium chunks (100 bits or so): command line interface,
Large chunks (1000 bits or more): program (batch) interface.

In any reasonably sophisticated system all three are needed, just as their metaphors are in natural language: phrasebook, dialogue, letter writing. All have their specific purposes, strengths and problems. To take the command line interface as a basis is a compromise: it makes possible to branch out to either side and to provide a menu interface through a menu command, and batch programs through command procedures.

4.3.1   The command line interface

> *(Admiral) 'If there should be war tomorrow, will the US or the USSR win?'*
> *(Computer, after 5 minutes) 'Yes'*
> *(Admiral) 'Yes what?'*
> *(Computer, after another 5 minutes) 'Yes, Sir'*

The amount of information transmitted in a command line is comparable to that contained in a sentence of a natural language, so there is a ready-made metaphor: the sentence. A short discussion of analogies, similarities and dissimilarities is in order. First of all, there is a basic asymmetry: a machine may possess some artificial intelligence, but in distinction to the human partner it is not supposed to have a free will (except in Sci-Fi).

A traditional distinction is that computing languages must be more precise than natural languages. Natural languages rely on common sense and guesswork on part of the recipient of the message, and on occasional requests for clarification. In the case of immediate languages, this distinction loses its stringency. While common sense is annoyingly difficult to achieve in a machine, intelligible error messages and requests for clarification or additional information are certainly feasible.

In natural languages, the typical sentence has three parts: Subject, Verb, Object, often in this order. Although Subject-last is unusual, all six possible orderings occur as the preferred order in some language. Word order is correlated in a subtle way with other aspects of syntax, see W. P. Lehmann (1978). Analogous typology studies for computing languages do not exist, but similar forces seem to be at work there too.

In computer languages, the subject of the sentence usually is implied (the user or the computer), but the object often is split into two, so that a typical command sentence again has three components: Verb, Input, Output. Again, each of the six possible orderings seems to occur as the preferred order of some language. Possibly, word order plays an even more significant role than in natural languages. For example, automated help facilities, such as prompting for omitted items, work best in traditional languages if the verb comes first, but in object oriented languages they seem to work most smoothly if the input object is first.

At present, two main syntax types dominate. One is the traditional command language, whose preferred word order is

Verb–Input–Output

as in 'copy x y', with the option of having multiple inputs and outputs. The other type is function style syntax, which seems to favor the word order

> Output–Verb–Input

as in '$y = f(x)$', with the option of having multiple inputs, but only a single (possibly composite) output. The opposite order V–I–O, as in '$f(x) =: y$', occurs less frequently.

Some early utilities for the PDP-10 used the weird order V–O–I.

As the name 'command' implies, the verb usually is an imperative, and occasional questions are disguised as orders to tell the answer.

In natural languages, a sentence corresponds to some unit of thought. A 'unit of thought' is a Protean concept; from time to time a frequently used combination will be recast into a new conceptual unit. Keeping in touch with the metaphor, also each command line should correspond to a (more or less) self-contained unit of thought. This implies that the building blocks of a command language ought to be chosen very carefully, neither too small, nor too big, so that they do not interfere with human thinking patterns. One should provide powerful, intuitive commands, and one should try to keep their number as small as possible, within reason. A large number of commands sometimes is advertised as a virtue of a system, while in reality it is a vice. It puts an unnecessary load on the user's memory, and hampers fluent use of a language.

By the way, in accordance with the 'one thought – one sentence – one command – one line' philosophy, the language should not make it possible to pack several commands onto the same line. While continuation lines occasionally may be needed, one should try to avoid them, their mere length makes them overly error-prone.

Command languages are designed for immediate use. This causes them to have characteristics very different from the more familiar batch programming languages.

First, an immediate language must be very robust with regard to errors. A moderately experienced user enters about a line per minute, and maybe a third of the lines are erroneous, ranging from mere typos to errors of thought, often to be discovered only several lines later. Errors are more frequent than in batch programming. It is necessary to design the language so that it does not invite errors. There are a number of quantitative empirical studies on programming errors in batch languages, see for example Litecky and Davis (1976), and note in particular their suggestion to deal with three of the four most error-prone features of COBOL by a relaxation of punctuation rules.

It must provide reasonable error messages, and easy recovery from errors. Division by zero, exponent overflow, and the like, should produce some form of NaN ('Not a Number') rather than a halt of execution. User errors must never produce a crash or hang-up. These are much tougher standards than those prevalent in current operating systems (see for example Miller, Fredriksen and So 1990, who were able to produce crashes or hang-ups with about a quarter of the utilities they tested).

Error-resistance is a strong argument in favor of the classical command line syntax, and against function-style syntax. The problem with the latter lies in its capability for nesting. Nesting is nice because then the user does not have to think up names for temporaries. On the other hand, nesting usually leads to error-producing piles of parentheses – an intelligent editor can prevent errors of the syntax, but not of the semantics of bracketing. Nesting of functions with side effects is a recipe for disaster: users do not pay attention to the order of evaluation. Since traditional mathematical notation uses function style syntax, one should use it there, but avoid it otherwise.

Second, the language must provide carefully designed user-assistance. Whenever something can have a default, it must have one. The language must insert defaults for omitted arguments, or prompt for them, if none is available. More generally, if an operation looks sensible by analogy, it must be permissible. This eases the load on the user's memory.

Third, declarations do not work well. Because of memory limitations and lack of foresight, humans are unable to declare things in advance, before actually needing them. In compiled languages, the principal purpose of declarations is to furnish the compiler with information that otherwise would become available only at link or run time, so that it can produce more efficient code. Typically, when writing a program, the programmer will repeatedly jump back to the program header and add declarations. In immediate languages, jumping back is impossible, and what may look like an array declaration, in reality is an executable statement, dynamically creating a new array filled with default zeros. Type declarations are replaced by inheritance rules: the left hand side of an assignment inherits its type and its dimensions from the right hand side. Integers are a subset of the real numbers, and should be treated as such (no distinction between short and long, signed and unsigned, please!). Side-effects of the lack of declarations are that immediate languages need dynamic memory management and a carefully designed garbage collection, and that highly optimized compilers are not feasible.

Forth, multi-line compound statements, looping and branching all belong to the batch realm. In immediate use, the incidence of typos is too high to make them practicable, they would need debugging. Whenever possible, the language therefore must eliminate loops by making them implicit. In other words, command languages ought to be array oriented, with implicit looping over all elements of the array.

Fifth, immediate languages need some degree of object-orientedness, both in data and in procedure objects. The question is how much. Clearly, all objects must carry hidden but accessible information about their own structure – at the very least data type and dimensions in the case of data objects, sophisticated built-in help in case of procedure objects. Though, excessive object-orientedness does not assist the user. Hiding structure increases complexity, if the user must access the hidden parts.

Sixth, the ability to keep legible records is crucial. In word processing, the final document is all that is needed. Similarly, in batch programming, each program version, when fully debugged, supersedes its antecedents. Data analysis is different: just as with financial transactions, the correctness of the end product cannot be checked without inspecting the path leading to it. Moreover, 'what-if' types of analysis are very common: one repeats an analysis with minor changes. Thus, it is necessary to keep a legible trail not only suitable for compiling a report on the analysis, but also a trail that can be audited, edited and re-executed.

4.3.2   The menu interface

The menu or WIMP interfaces (Windows-Icons-Menus-Pointing) are indispensable in graphics applications: a single mouse-click transfers some 20 bits of coordinate information very effectively. Data analysis uses graphics as an essential and effective method to exchange true quantitative graphical information in *both* directions between humans and computers.

Somewhat perversely, the most common among the WIMP interfaces, the so-called graphical user interfaces (GUIs) have nothing to do with graphics at all. The tiny amount of information (1-4 bits) transmitted in most elementary menu interactions is both an advantage and a liability. The main advantage is that a well-designed menu system offers tight guidance and can prevent user errors by not even offering the opportunity to commit them. This way, it can greatly assist the inexperienced or occasional user in doing an infrequent, standardized task.

Navigating a tree of menus can be very frustrating to a newcomer who easily gets lost in a labyrinth of branches, but also to an experienced user who knows where to go but cannot get there quickly. Few menu systems offer any means to retrace one's steps. This can be most aggravating for example when one is installing a major piece of software and overlooks the necessity to set some option. Typically, one then has to start from scratch, loses concentration, blunders again, and so on.

The greatest liability of menu systems is that one cannot easily automate repetitive steps. The maxim of early UNIX "Don't insist on interactive input" (UNIX 1978, p.1902) should not have been forgotten!

Menu systems generally are rigid and not user-extensible. Attempts to provide at least some small extensions through macros are pathetic. The paper by Rawal (1992) is typical and may serve as an example. It describes a facility to capture a series of mouse and keyboard events into a macro and then to bind that macro to a key press event. Such extensions negate the advantages and strengths of the menu and of the mouse. The mere fact that somebody feels a need for such a wretched facility shows that there is something rotten with WIMP-based systems.

In applications like data analysis an almost equal liability is the lack of legible, editable and executable action trails. The problem lies with the smallness of the chunks of information, so by necessity there is extreme context-sensitivity. This makes it difficult to combine the small chunks into humanly intelligible records. The examples of macros contained in Rawal's paper illustrate the problem: each mouse-click is expanded into a line of text, but still, the meaning of that click is far from evident.

Saving an executable trail is easy, the problem is with editing it. In the mid-1980's we (Russell Almond, David Donoho, Peter Kempthorne, Ernesto Ramos, Len Roseman, Mathis Thoma and myself) were making a few movies to illustrate the use of high-interaction graphics in data analysis. After some trial and error we did this by capturing a complete trail of mouse and keyboard events. Then we edited the trail, and finally shot a movie, frame-by-frame, by running the edited trail overnight. Mathis Thoma wrote a sophisticated program to facilitate the editing part. The tricky part with editing was that one had to extract the intentions of the user and then reconstruct a new trail. Informally, a typical editing instruction might have been: 'Rotate the scene from position A to position B as in the original trail, but do it more smoothly, and in 7 instead of 5 seconds, so that the spoken commentary can be fitted in'.

The information transmitted in an elementary menu interaction typically is smaller than what would correspond to a natural unit of thought. An interesting consequence of this is that keyboard-based menu languages tend to develop into a weird kind of command language: short key sequences are used as units, without glancing at the menu between hitting the keys. Thus, the key sequence 'F10 q q Ins' may take over the role of the word 'quit' in a traditional command language, and 'F10 p f Ins' that of the word 'print' (example from Lotus Manuscript)!

4.3.3 The batch interface and programming environments

In traditional batch style programming, the user writes (or edits) a program, then compiles, links and executes it. Usually, many cycles through these activities are needed for testing and tuning before the program can be used in production. Traditional batch programming *per se* clearly is needed and used in the final stages of prototyping and for preparing a production system. This type of operation is traditional and well-understood. It usually is done under control of a *make*-like facility: after a change, the entire collection of modules is re-linked and re-started from scratch.

Here we are primarily concerned with the much less familiar batch-style paradigm in immediate computing, where computation is continued after a change without necessarily re-starting from scratch. It was introduced for program development by the AI community and was implemented primarily on SMALLTALK and LISP machines. In sophisticated applications like data analysis, this paradigm takes on a special flavor going beyond that of mere program development, since it is tightly coupled with actual, applied production work. The main difference is that the role of the programmer is taken over by that of the data analyst, and the focus is not on the program but on the data. At least in the research and development phase, what is programmed is dictated by *ad hoc* needs derived from the data at hand, rather than from program specifications: an exploratory session rarely follows the planned course more than 15-30 minutes, then one begins to improvise. Typically, one will work in immediate mode, dispatching pre-defined or previously prepared modules for immediate execution. But often, it becomes necessary to extend the system on the fly by writing short program modules, or by modifying existing modules. All those modules must be able to work together. Some of them are there to stay, for example there must be a large and stable kernel for doing standard tasks. Some ephemeral ones will be improvised for the task at hand. Very often, one will extract a sequence of command lines that have been executed in immediate mode and transform them into a new command module by adding some branching or looping structure and some lines mediating data exchange. There must be a rigorous protocol for exchanging data and for documenting new modules, and there must be safeguards against a buggy module tearing down the rest.

Such an approach clearly can be effective only if there is a decent programming environment. By this, we mean an environment in which the user can create, edit and debug a program module in the midst of a session, without losing intermediate results generated earlier in the same session. The traditional compile-link-execute cycle must either take negligible time, or be completely eliminated. None of the classical programming languages (FORTRAN, PASCAL, C, . . .) can provide such an environment; apart from the AI languages LISP and SMALLTALK, only APL and a few versions of BASIC do.

The traditional distinction between operating system and programming language vanishes in a good programming environment. The functions of the two therefore often are combined into a single language; examples include operating systems based on LISP, SMALLTALK or BASIC. Some data analysis systems have been built directly on top of such environments, see e.g. Stuetzle (1987). Though, these efforts have now faded away together with their special purpose hardware.

Technically, all these approaches use the immediate batch interface: a program module of an arbitrary size is dispatched out of the editor for execution. In practice however, it tends to coalesce with the command language approach: the complexity of frequently used larger modules is hidden behind smaller modules ('commands') invoking them. Conversely, command languages must be extensible on the fly, hence it is necessary to build a programming environment into the command language.

In any case, in an extensible language, new commands are batch constructs. They can be added either as interpreted command procedures ('soft commands') or as compiled code to be linked with the overall system ('hard commands'), or as separate main programs, to be invoked as a child process by a small command procedure responsible for mediating the data interchange. If command procedures are interpreted, looping is slow, but in data analysis this efficiency problem arises surprisingly rarely: with proper programming, most jobs can be done through fast, implicit looping over all array elements. Recursive calculations with scalars are among the few cases where this will not work.

From the linguistic point of view it does not matter whether the modules are interpreted, compiled, or subjected to run-time compilation, this is a matter of implementation.

4.3.4   Some personal experiences

Over the past 25 years I have used our own data analysis language (ISP) extensively for various purposes. The most demanding among them were those where the results of the analysis were intended for inclusion in a paper, and where one had to produce tables and publication-grade figures, the latter as encapsulated Postscript files.

As a rule, my analyses mostly began with writing a few soft commands and using them in immediate mode. Sooner or later, segments of session transcripts were then pasted together into larger programs to be used in batch mode. These programs then were fine-tuned, using the programming environment facility of ISP to edit and resubmit them in quick succession. Programs tended to grow until a single large program would produce all the displayed material to be included in some paper. The main reason for using batch mode was that publication-grade data graphics needs rather delicate fine tuning, and you always need a few more turn-arounds than you anticipate. More than once I had to re-produce an entire set of pictures by running a new data set through the program – for example because an additional year of geophysical data had become available.

The main change over these 25 years was a very considerable increase of computer performance, in fact so much that in my case it usually became feasible to encapsulate an entire analysis in a program with a small number of interactive switches, and to rerun it from scratch if the need should arise. Session transcripts served as the basis for programming but were no longer necessary to document the analysis – that task was taken over by a program written in a legible command language.

4.4   MISCELLANEOUS ISSUES

4.4.1   On building blocks

> *Simple, elegant solutions are more effective, but they are harder to find than complex ones, and they require more time.*   (N. Wirth, 1985)

The choice of the right building blocks is absolutely essential in any computing language, both in the core of the language and in its extensions, where they take the form of library procedures. If the blocks are too small, that is, smaller than natural units of thought, we end up with an annoying kind of assembly language programming. If they are too large, they become strait-jackets, impeding flexibility. Side effect can become awkward or fatal. For example, the X11 library (see X Window 1993) contains about 20 routines for dealing with the event queue. Most of them create annoying side effects by combining at least two unrelated actions. For example, there is no function that simply checks whether a particular event has happened, without waiting, and without modifying the contents of the queue. Five or six well-chosen simpler functions would do a better job with less aggravation.

Once a building block has been included in a publicly distributed library, it is almost impossible to get rid of it without creating compatibility problems. The conclusion is: in case of doubt, leave it out.

4.4.2   On the scope of names

The *scope of names* is a notorious open question in programming languages. A safe and clean solution is to treat all names occurring in a procedure as local to that procedure, except if they are imported or exported in an explicit fashion. However, this turns out to be very bothersome in a language intended for immediate use, because one then may have to spell out too many names. An alternative solution (going back to ALGOL) is to treat all names as global, unless they have explicitly been made local. This permits selective hiding of information. The problem is, of course, that a forgotten local declaration can lead to nasty errors through an inadvertent modification of a global variable.

4.4.3   On notation

A computing language should be easy to write and to read; ease of reading is the more important of the two. A program, or a session transcript, must be easy to scan, even by somebody who is only moderately proficient in the language. Notational details can make a big difference, and it is important to keep a wise balance between terseness and redundancy.

A basic decision is on the kind of syntax the language should follow. Operating system shells suggest a *classical command language*. Traditional mathematical notation suggests a *function style* approach. In the case of ISP, we went for a compromise: use both, but restrict the latter to built-in simple functions. A dogmatic insistence on purity only hampers the average user who is not a computer science professional.

The reason against using function style throughout has to do with its ability for nesting: nesting of complicated functions is error-prone in immediate mode, and it is difficult to provide graceful user assistance (e.g. prompts for omitted or erroneous arguments), if also the errors can be nested.

On the other hand, one should follow established models for expressing mathematical functions. After all, traditional mathematical notation has undergone several centuries of natural selection. For example, $x * (y + z)$ is easier to read than the reverse Polish version $x\ y\ z\ +\ *$, which, however, is easier to use on a calculator.

Assignments are customarily written in the form $x = \cos(t)$. The alternatives $x := \cos(t)$ or $\cos(t) =: x$ are at a disadvantage, partly because one of $:=$ and $=:$ is enough (one should generally avoid offering equivalent alternatives to solve the same problem), and partly because assignments are so frequent that typing a digraph is a nuisance. Symbols (e.g. arrows) that cannot be found on the usual keyboards should be avoided.

Infix notation, with the operator between the two operands, should remain reserved to the traditional cases, that is, to the four arithmetic operators $+$, $-$, $*$, $/$ and exponentiation $**$ (or $\char`\^$), the relational operators $<$, $<=$, $>$, $>=$, $==$, $!=$, and the logical operators $\&$, $|$.

In case of doubt, one should use self-explanatory notation. For example, the meaning of the expression

$$z = \text{ if } a > b \text{ then } a \text{ else } b$$

is reasonably easy to figure out, namely $z = \max(a, b)$, while a C-style

$$z = (a > b) \text{ ? } a : b;$$

is baffling to the uninitiated.

4.4.4   Book-keeping problems

With large multi-session data analyses, book-keeping problems get severe, comparable to those of large multi-person software projects. It is not sufficient to keep readable, editable and executable records, it must also be possible to retrieve specific information from those records. For example, if one learns about a serious error in a data set or a program module, it is necessary to find out which parts of the analysis are affected and must be repeated. This is considerably trickier than the mere use of a *make*-facility and involves AI-type programming. A system for script analysis must for example know about the syntax and semantics of the underlying data analysis language. For a brief discussion of the issues see Huber (1986b). After some preliminary work by W. Nugent a prototype "Lab Assistant" was subsequently programmed in LISP by W. Vach (1987). His system was able to hunt down forward and backward logical dependencies, and to put up a warning signal, where a dependency might have been introduced by an invisible outside intervention (e.g. by the user keying in a literal number read off the screen). To talk about languages: since we were aware that this was an AI project, we felt that we had to use an AI language. We first thought of PROLOG, but then we settled on LISP, because it was stabler. However, in retrospect it would have been preferable to expand the data analysis language ISP and its text handling capabilities to the point that it could handle its own session analysis, and to use the information already built into it rather than to duplicate its parser, syntax analyzer and semantics in LISP. Otherwise, it simply is not possible to keep a session analysis system in step with an extensible and developing data analysis system.

Thanks to these prototypes, we realized that "Lab Assistants" not only were trickier to build than we had anticipated, but that the presence of invisible dependencies would make them more bothersome to operate and therefore less useful than we had thought. Fortunately, with improved computer performance the need for sophisticated instruments of this type began to fade away. In most cases it would be simpler to repeat the entire analysis from scratch, rather than to identify and skip the parts that would not need to be repeated. Still, such a repetition is only possible if one keeps invisible outside interventions to a minimum and takes care to save a complete executable trail of the analysis. See also Section 4.3.4 and the doubts expressed in Section 2.6.4, whether it would be possible to design a successful Lab Assistant system serving more than one master.

4.5   REQUIREMENTS FOR A GENERAL PURPOSE IMMEDIATE LANGUAGE

A system suitable for statistics and interactive data analysis must satisfy all the requirements outlined so far for a general purpose immediate language. The more integration there is, the better. We can summarize the language requirements as follows:

(1) *Simplicity.* It must have a simple command line syntax. The language must be easily writable and readable, and it should not be error-prone. Hence it should put little reliance on special symbols and punctuation (except for visual cues).

(2) *Extensibility.* It must be extensible and moderately object oriented, with a completely general language kernel, and with the same syntax for hard and soft commands (i.e. built-in commands and command procedures). It must offer a full interactive programming environment.

(3) *Records.* It must have full record keeping capabilities.

(4) *Help.* There must be decent, automatic on-line help, with prompts for omitted arguments, rather than mere error messages.

(5) *Multimedia.* There must be various primitives not only for numeric, but also for text operations, and possibly for voice and images, too.

(6) *Openness.* It must offer interactive and programmed access to the outside world without losing the current environment.

(7) *I/O.* In particular, it must offer flexible and efficient I/O facilities that can be adapted quickly and easily to handle arbitrary data structures, binary and otherwise.

(8) *Missingness.* There must be a sensible default treatment of missing values.

(9) *Tools.* It must offer all standard functions, linear algebra, some carefully selected basic statistical and data base management tools, random numbers, and the main probability distributions.

(10) *Graphics.* It must have tightly integrated general purpose tools for high-level, high-interaction graphics and menu construction.

The only items specific to statistics occur in items (7) to (10).

More advanced statistical tools do not belong into the language itself, but into the 'literature' written in that language, and into utilities to be accessed through (6). This way, the linguistic demands made on the user are kept to a minimum: command procedures are written in the general purpose data analysis language, while programs accessed through (6) can be written in the user's favorite language, whatever that is. The gray zone between language core and literature demands constant attention, if it is not to grow out of hand.

Some of our experiences with ISP relative to large-scale, non-standard data analysis deserve to be detailed.

Graphics. ISP was the first system to add and integrate high-interaction graphics. The graphics part was soon copied by other systems (e.g. MacSpin), but the need for integration usually was neglected or misunderstood. Without such integration – in particular without the ability to create complex subset structures and to exchange them forth and back between graphics and the general purpose, non-graphical data analysis language – interactive data graphics is little more than a video game.

Exploration graphics must be simple and fast to use, that is, it must rely mostly on defaults.

Presentation graphics in data analysis tends to have rather unpredictable and high demands. The standard graphics packages (Harvard Graphics, Photoshop, etc.) are not up to the task, especially not, if dozens or hundreds of similar pictures must be produced in the same uniform, externally specified format. In such cases, our users resorted to writing Postscript files via ISP procedures.

Fast prototyping and production systems. With a flexible, integrated system one can prototype complicated, non-standard, special purpose applications within days or weeks. The prototype is developed through interaction between the prototype designer and the end-user, while the latter applies it to his or her actual data. This way, the user can begin to work with the prototype very early, which permits to correct mis-specifications and design errors before they become too costly. Since the prototype is patched together from generic building blocks, the seams between the blocks may show, but it is not a mere mock-up, it flies. At the end of the prototyping phase it sometimes may even be judged good enough to do the actual production work.

However, production work with large real data means that the demands on the systems increase. A fortunate aspect of array languages is that efficiency of interpretation relative to compilation improves with array size.

An experience derived from this type of work is: if the final production system is to be operated through menus, then the menu tree cannot be specified in advance, it must be constructed on the basis of the application, and *together* with it. In order that this could be done, a generic menu command became indispensable.

Another repeated experience derived from fast prototyping is that the language must be able to handle sophisticated *data base management* tasks. Large real-life problems always require a combination of data base management and data analysis. The larger the data sets, the more complicated their structure tends to be. Neither data base management systems nor traditional statistics packages are up to the task.

Concluding remarks

A language for data analysis must not only serve for *doing* data analyses, but also for *describing* them in an unambiguous, humanly legible form.

The core of a language for data analysis is completely general and contains negligibly few items *specific* to data analysis.

The combined significance of these two remarks is that any such language has the potential of developing into a *koiné*, a common standard computer language. If well implemented, it could easily take over both the role of an operating system shell and of a very high level programming language. In view of Naur's comments, some of which have been quoted in Section 4.2.2, my optimism is muted. But it might happen if and when the software towers of Babel erected on top of the currently fashionable operating systems should crumble under their own weight.

CHAPTER 5

APPROXIMATE MODELS

This chapter is based on a talk I gave at the International Conference on *Goodness-of-Fit Tests and Model Validity*, at Paris, May 29–31, 2000, commemorating the centennial anniversary of the landmark paper by Karl Pearson (1900) on chi-square goodness-of-fit test (see Huber 2002)[1].

5.1 MODELS

The anniversary of Karl Pearson's paper offers a timely opportunity for a digression and to discuss the role of models in contemporary and future statistics, and the assessment of adequacy of their fit, in a somewhat broader context, stressing necessary changes in philosophy rather than the technical nitty-gritty they involve. The present chapter elaborates on what I had tentatively named "postmodern robustness" in Huber (1996b, final section):

[1] ©With kind permission from Springer Science+Business Media.

Data Analysis: What Can Be Learned From the Past 50 Years. By Peter J. Huber
Copyright © 2011 John Wiley & Sons, Inc.

The current trend toward ever-larger computer-collected and computer-managed data bases poses interesting challenges to statistics and data analysis in general. Most of these challenges are particularly pertinent to diagnostics and robustness. The data sets are not only getting larger, but also are more complexly structured. [...] Exact theories become both impossible and unmanageable. In any case, models never are exactly true, and for sufficiently large samples they will be rejected by statistical goodness-of-fit tests. This poses some rather novel problems of robust model selection, and we must learn to live with crude models (robustness with regard to systematic, but uninteresting, errors in the model). [...] It appears that none of the above problems will be amenable to a treatment through theorems and proofs. They will have to be attacked by heuristics and judgment, and by alternative "what-if" analyses. But clearly all this belongs to robustness in a wide sense. Does this signal the beginning of an era of "postmodern" robustness?

Karl Pearson's pioneering 1900 paper had been a first step. He had been concerned exclusively with distributional models, and with global tests of goodness-of-fit. He had disregarded problems caused by models containing free parameters: how to estimate such parameters, and how to adjust the count of the number of degrees of freedom. Corresponding improvements then were achieved by Fisher and others. Though, the basic paradigm of distributional models remained in force and still forms the prevalent mental framework for statistical modeling. For example, the classical texts in theoretical statistics, such as Cox and Hinkley (1974) or Lehmann (1986), discuss goodness-of-fit tests only in the context of distributional models. Apart from this, it is curious how current statistical parlance categorizes models into classes – such as "linear models" or "generalized linear models" or "hierarchical models". The reason is of course that each such class permits a specific formal statistical theory and analysis, or, expressing it the other way around: each class constitutes the natural scope of a particular theoretical framework. In my opinion, to be elaborated below, such categorizations can have the undesirable consequence that one is fitting the data into pre-ordained model classes.

We now have a decisive advantage over Pearson: whenever there is an arbitrarily complex model and a test statistic, we can, at least in principle, determine the distribution of that statistic under the null hypothesis to an arbitrary degree of accuracy with the help of simulation. It is this advantage which allows us to concentrate on the conceptual aspects, in particular on the traditionally suppressed problems posed by partial or otherwise inaccurate models.

The word "model" has a bewilderingly wide semantic range, from tiny trains to long-legged girls. Though, it hides a simple common feature: a model is a representation of the essential aspects of some real thing in an idealized, exemplary form, ignoring the inessential ones. The header of this chapter is an intentional pleonasm: by definition, a model is not an exact counterpart of the real thing, but a judicious approximation. Mathematical statisticians, being concerned with the pure ideas, sometimes tend to forget this, despite strong admonitions to the contrary, such as by McCullagh and Nelder (1983, p.6): "all models are wrong". What is considered essential of course depends on the current viewpoint – the same thing may need to be modeled in various different ways. Progress in science usually occurs through thinking in models, they help to separate the essential from the inessential.

A general discussion of the role mathematical models have played in science should help to clarify the issues. The models can be qualitative or quantitative, theoretical or empirical, causal or phenomenological, deterministic or stochastic, and very often are a mixture of all. The historical development of the very first non-trivial mathematical models, namely those for planetary motion, illustrates some salient points most nicely, namely (1) the interplay between conceptual/qualitative and phenomenological/quantitative models, (2) the discontinuous jumps from a model class to the next, and (3) the importance of pinpointing the discrepancy rather than merely establishing the existence of discrepancies by a global test of goodness-of-fit.

Early Greek scientists, foremost among them Anaxagoras (ca. 450 BC), had tried to explain the irregular motions of the planets by a "stochastic model", namely by the random action of vortices. This model did not really explain anything, it just offered a convenient excuse for their inability to understand what was going on in the skies. In the 4th century BC, Eudoxos then invented an incredibly clever qualitative model. He managed to explain the puzzling retrograde motion of the planets deterministically in accordance with the philosophical theory that celestial motions ought to be circular and uniform. For each planet, he needed four concentric spheres attached to each other, all rotating uniformly. Quantitatively, the model was not too good. In particular, even after several improvements, it could not possibly explain the varying luminosity of the planets, because in this model their distances from the earth remained constant. About the same time the Babylonians devised empirical models, apparently devoid of any philosophical underpinning: they managed to describe the motions of the moon and the planets phenomenologically through arithmetic schemes involving additive superposition of piecewise linear functions. Around AD 130, Ptolemy then constructed an even cleverer model than Eudoxos. He retained the politically correct uniform circular motion, but the circles were no longer concentric. With minor adjustments this type of model withstood the tests of observational astronomy for almost 1500 years, until Kepler, with the superior observations of Tycho Brahe at his disposal, found a misfit of merely $8'$ (a quarter of the apparent diameter of sun or moon, and just barely above observational accuracy) in the motion of a single planet (Mars). He devised a fundamentally new model, replacing the uniform circular by elliptical motions. The laws describing his model later formed the essential basis for Newton's theory of gravitation.

For us, Kepler's step is the most interesting and relevant. First, we note that the geocentric Ptolemaic and the heliocentric Copernican models phenomenologically are equivalent – both belong to the class of epicyclic models, and with properly adjusted parameters the Ptolemaic model renders the phenomena, as seen from the earth, absolutely identically to the Copernican one. But the conceptual step to heliocentricity was a most crucial inducement for Kepler's invention. Second, Kepler's own account shows how difficult it is to overcome modeling prejudices. It required the genius of Kepler to jump over the shadow cast forward by 1500 years of epicyclic modeling. For us, to resist and overcome the temptation to squeeze our data into a preconceived but wrong model class – by piling up a few more uninterpretable parameters – may require comparable independence of mind. I think that here we have a warning tale about the dangers of categorizing models into narrowly specified classes.

5.2 BAYESIAN MODELING

In the years following the paper by Geman and Geman (1984), modeling activity in the statistical literature has concentrated heavily on Bayesian modeling and on Markov Chain Monte Carlo methods; see the survey by Besag et al. (1995). There are points of close contact of this activity with the present chapter, namely the common concern with complex models and simulation.

Modeling issues provide one of the neatest discriminations between the relative strengths and weaknesses of the Bayesian and frequentist approaches. The two complement each other in a most welcome fashion. The last paragraph of this section contains an illustrative example. The frequentist-Pearsonian approach has difficulties with the comparative assessment of competing models. P-values, when abused for that purpose, are awkward and unintuitive, to say the least, while the Bayesian approach provides a streamlined mechanism for quantitative comparisons through posterior probabilities.

On the other hand, Bayesian statistics lacks a mechanism for assessing goodness-of-fit in absolute terms. The crux of the matter, alluded to but not elaborated in the excellent discussion of scientific learning and statistical inference by George Box (1980, see p. 383f.), is as follows. Within orthodox Bayesian statistics, we cannot even address the question whether a model M_i, under consideration at stage i of the investigation, is consonant with the data y. For that, we either must step outside of the Bayes framework and in frequentist-Pearsonian manner perform a goodness-of-fit test of the model against an unspecified alternative (an idea to which Box, being a Bayesian, but not a dogmatic one, obviously did not object), or else apply "tentative overfitting" procedures. The latter are based on the unwarranted presumption that by throwing in a few additional parameters one can obtain a perfectly fitting model. But how and where to insert those additional parameters often is far from obvious, unless one is aided by hindsight and is able to abandon one's tacit prejudices about the class of models. Remember that Kepler rejected the epicyclic Ptolemaic/Copernican models because he could not obtain an adequate fit within that class.

The tutorial on Bayesian model averaging by Hoeting et al. (1999), through the very tentativeness of their remarks on the choice of the class of models over which to average (what are "models supported by the data"?) further highlights the problems raised by model adequacy. To put the spotlight on the central problem, we consider the special case of a small finite number of discrete models (such that the fit cannot be improved by throwing in additional parameters). Before we can meaningfully average, or assign meaningful relative probabilities to the different models by Bayesian methods, we ought to check whether the models collectively make sense. A mere verification by a goodness-of-fit test that the class contains models not rejected by such a test, is not good enough. Averaging over a class of spurious models is not better and safer than treating a single non-rejected model as correct – see the next section – and maybe even worse, because it obscures the underlying problem.

Only in very exceptional cases one can check appropriateness of a class of models by a reverse application of a goodness-of-fit test, namely by testing and rejecting the hypothesis that all models under consideration are wrong. One of the rare cases where such a "badness-of-fit" test works occurs in Huber et al. (1982). There, the problem had been to date a certain Old Babylonian king with the help of astronomical observations. Venus observations from his reign gave k possible dates, with k being anywhere between 4 and 20, depending on the time range one considers to be historically feasible, with 20 being overly generous. Venus phenomena show approximate periodicities and therefore cannot give unique dates. That is, we have here a case of k distinct, discrete models, as alluded to in the preceding paragraph. Unfortunately, it was not entirely certain whether the Venus observations really came from the reign of the king in question, so the k dates possibly might have been all wrong. Independent month-length data – the lengths of lunar months alternate quite irregularly between 29 and 30 days – from economic texts dating to his reign and to that of his predecessor then were compared against calculation. The expected agreement rate for randomly chosen wrong chronologies (53%) is securely known from astronomical theory, while the higher rate for a correct chronology (67%) is based on Late Babylonian control material and is less secure. The month-lengths were used in Neyman-Pearson fashion to test, and reject on the 5% level, the hypothesis that all $k = 20$ model dates were wrong (or, with $k = 4$, on the 1% level). For that, only the securely known agreement rate for wrong chronologies is required. Logically speaking, such a test would tell us that one of the k models was correct, but not which one. Finally, a Bayesian argument was used with all available data to assign relative likelihoods (posterior probabilities for a uniform prior) to the k possible dates. The winning date garnered 93% of the posterior probability.

5.3 MATHEMATICAL STATISTICS AND APPROXIMATE MODELS

Karl Pearson, as can be seen from a few remarks scattered through his 1900 paper, had been perfectly aware that his distributional models were approximations. For example, he notes that a model fitting well for a particular sample size would fit for smaller, but not necessarily for larger samples.

After Karl Pearson, under the influence of Fisher and others, mathematical statistics developed a particular *modus operandi*: take a simple, idealized model, create an optimal procedure for this model, and then apply the procedure to the real situation, either ignoring deviations altogether, or invoking a vague continuity principle.[2] By 1960, this approach began to run out of steam: on one hand, one began to run out of manageable simple models, on the other hand, one realized that the continuity principle did not necessarily apply: optimization at an idealized model might lead to procedures that were unstable under minor deviations. The robustness paradigm – explicitly permitting small deviations from the idealized model when optimizing – carried only a few steps further.

Statisticians, as a rule, pay little attention to what happens after a statistical test has been performed. If a goodness-of-fit tests rejects a model, we are left with many alternative actions. Perhaps we do nothing (and continue to live with a model certified to be inaccurate). Perhaps we tinker with the model by adding and adjusting a few more features (and thereby destroy its conceptual simplicity). Or we switch to an entirely different model, maybe one based on a different theory, or maybe, in the absence of such a theory, to a purely phenomenological one.

In the opposite case, if a goodness-of-fit test does not reject a model, statisticians have become accustomed to act as if it were true. Of course this is logically inadmissible, even more so if with McCullagh and Nelder one believes that all models are wrong *a priori*. The data analytic attitude makes more sense, namely to use such tests in moderately sized samples merely as warning signs against over-interpretation: renounce attempts at interpreting deviances from the model if a goodness-of-fit test (with a much higher than usual level) fails to reject the model.

---

[2]There has been curiously little concern in the statistical literature with the optimality of goodness-of-fit tests themselves (i.e. of χ^2- and F-tests), and it has been relegated to notes, such as by Lehmann (1986), 428-429.

Moreover, treating a model that has not been rejected as correct can be misleading and dangerous. Perhaps this is the main lesson we have learned from robustness. A few illustrations follow.

(1) Model-optimality versus fit-optimality. Distributional goodness-of-fit tests typically are based on the value of the distance measured between the empirical distribution F_n and the model distribution. To fix the idea, let us take the Kolmogorov distance $K(\mu, \sigma)$, which makes exposition simpler (but exactly the same arguments and results apply to a distance based on the χ^2 test statistic):

$$K(\hat{\mu}, \hat{\sigma}) = \sup_x \left| F_n(x) - \Phi\left(\frac{x - \hat{\mu}}{\hat{\sigma}}\right) \right| \tag{5.1}$$

where $\hat{\mu}$, $\hat{\sigma}$ are either the traditional model-optimal estimates:
  (a) $\hat{\mu}$, $\hat{\sigma}$ = ML estimate for model Φ
or the fit-optimal estimates:
  (b) $\hat{\mu}$, $\hat{\sigma}$ = minimizer$_{\mu,\sigma}$ $K(\mu, \sigma)$.

In case (a), the test sometimes correctly rejects the hypothesis of normality for the wrong reason, namely if an outlier inflates the ML estimate for σ, even though the fit, in terms of minimized distance (b), is good. The disturbing fact is that the traditional recommendation, namely to estimate the unknown free parameters by any asymptotically efficient estimate, *viz.* either maximum likelihood or minimum χ^2, may have very different consequences depending on which estimate one chooses.

(2) Minimax robust estimates. Observational errors in most cases are excellently modeled by the normal distribution, if we make allowance for occasional gross errors (which may be of different origin). If we formalize this in terms of a contamination model, then the normal part represents the essential aspects of the observational process in an idealized form, the contamination part merely models some disturbance factors unrelated to the quantities of interest. But model-optimal estimates for the idealized model, e.g. the sample mean, are unstable under small deviations in the tails of the distribution, while robust estimates, such as judiciously chosen M-estimates, offer stability but lose very little efficiency at the normal model.

Typically, we are not interested in estimating the parameters of the contaminant, or to test it for goodness-of-fit, and mostly not even able, given the available sample size. Note that a least favorable distribution is not intended to model the underlying situation (even though it may approximate the true error distribution better than a normal model), its purpose is to provide a ML estimate that is minimax robust.

(3) Linear fit. Assume that you want to fit a straight line to an approximately linear function that can be observed with errors in a certain interval $[a, b]$. Assume that the goal of the fit is to minimize the integrated mean square deviation between the true, approximately linear function and an idealized straight line. A model-optimal design will put one half of the observations at each of the endpoints of the interval. A fit-optimal design will distribute the observations roughly uniformly over the interval $[a, b]$. The unexpected and surprising fact is that subliminal deviations of the function from a straight line (i.e. deviations too small to be detected by goodness-of-fit tests) may suffice to make the fit-optimal design superior to the model-optimal design (cf. Huber (1975b)).

In short: for the sake of robustness, we may sometimes prefer a (slightly) wrong to a (possibly) right model.

5.4 STATISTICAL SIGNIFICANCE AND PHYSICAL RELEVANCE

It is a truism that for a sufficiently large sample size any model ultimately will be rejected. But this is merely a symptom of a more serious underlying problem, involving the difference between statistical significance and physical relevance. Since this is a problem of science rather than one of statistics, statisticians unfortunately tend to ignore it. The following example is based on an actual consulting case.

Example: A computer program used by physicists for analyzing certain experiments automatically plotted the empirical spectrum on top of the theoretical spectrum, and in addition it offered a verbal assessment of the goodness-of-fit (based on a χ^2 statistic). Perplexingly, if the fit was visually perfect, it would often be assessed by the program as poor, while a poor visual fit often would be assessed as good. At first, the physicists suspected a programming error in the statistical calculations, but the reason was: the experiments had widely different signal-to-noise ratios, and if the random observational errors were large, the test would not reach statistical significance, and if they were small, the test would reject because of systematic errors (either in the models or in the observations), too small to show up in the plots and probably also too small to be physically relevant.

With increasing data sizes, the paradoxical situation of this example may occur even more often: if a global goodness-of-fit test does not reject, then the observations are too noisy to be useful, and if it rejects, the decision whether or not to accept the model involves a delicate judgment of relevance rather than of statistical significance.

In short, in real-life situations the interpretation of the results of goodness-of-fit tests must rely on judgment of content rather than on P-values. For a traditional mathematical statistician, the implied primacy of judgment over mathematical proof and over statistical significance clearly goes against the grain. John Tukey once said, discussing the future of data analysis (1962, p.13): "The most important maxim for data analysts to heed, and one which many statisticians seem to have shunned, is this: 'Far better an approximate answer to the right question, which is often vague, than an exact answer to the wrong question, which always can be made precise.' " Analogous maxims apply to modeling: a crude answer derived in a messy way from an approximately right model is far better than an exact answer from a wrong but tractable model.

5.5 JUDICIOUS USE OF A WRONG MODEL

A striking example of the judicious use of a wrong model that deserves our attention for methodological reasons is discussed by Tukey (1965, p. 269-270). It concerns the surprising discovery by Munk and Snodgrass (1957) of ocean waves reaching the Pacific Coast of the United States from very distant sources.

Many accounts of spectrum analysis emphasize the usual hypothesis of stationarity; after all, spectrum analysis is based on the theory of stationary stochastic processes. Some go so far as to leave the impression that it is only applicable to stationary time series. Quite contrary to this impression, the remote waves were only discovered because of their nonstationarity – and because this nonstationarity was studied by spectrum analysis.

The most effective method of measuring longer-period waves on the ocean near land, without interference from the shorter-period waves that make up locally generated surf, is to place a pressure recorder on the bottom at an appropriate depth and location. The depth determines which wavelengths will be recorded, since the penetration of pressure fluctuations depends exponentially on the ratio of depth to wavelength. The location determines the exposure to waves reflected from the shore. Munk and Snodgrass operated such a recorder continuously on the seaward side of Guadeloupe Island, 200 miles off the coast of Mexico. They divided their records into four-hour sections and submitted the sections to numerical spectrum analysis. Along with much larger features that they had come to expect and understand, they noted a small peak (located at the low-frequency side of a very much larger peak). This peak appeared in successive four-hour records in almost the same place. Actually its frequency moved steadily toward higher frequencies at a measurable rate.

This nonstationarity applied a definite label to this group of waves. Surface waves on water are dispersive, by which is meant that longer waves travel at a different velocity, actually faster, than shorter ones. Accordingly, if a distant source of waves exists for a limited time, lower frequency waves will arrive first, followed by the shorter, higher-frequency ones. The more distant the source, the slower the change. Measurement of nonstationarity thus measures the source distance.

The dimensions of this group of waves were found to be as follows:

height:	about 1 millimeter
length:	about 1 kilometer
source distance:	about 14000 kilometers.

There was only one plausible location for the source of waves at this distance: the Indian Ocean. At this point, source location was a matter of plausible inference. But subsequent further research, including cross-spectrum analysis of measurements taken by three recorders placed at the corners of a triangle, then permitted to trace the sources of these waves to heavy storms at the right time and place.

As Tukey pointed out: The whole history of remote waves is dependent on the use of conventional methods of spectrum analysis, supposedly assuming stationarity, to study, measure and utilize the nonstationarity of the arriving waves.

5.6 COMPOSITE MODELS

The problems considered and solved by mathematical statisticians are model problems themselves – they idealize exemplary situations. Unfortunately, these problems and their solutions often are too simple to be directly applicable and then ought to be regarded as mere guidelines. Applied statisticians and data analysts today are confronted by increasingly more massive and more complex real-life data, requiring ever more complex models. Such complex models then must be pieced together from simpler parts.

Composite models facilitate a conceptual separation of the model into important and less important parts, and they also make it easier to locate where the deviations from the model occur. Usually, only some parts of the model are of interest, while other parts are a mere nuisance. Sometimes, in traditional fashion, we may want to test for the presence or absence (perhaps more accurately: lack of importance) of certain features by not including them in the model. But we now have to go beyond. We may be willing to condone a lack of fit in an irrelevant part, provided it does not seriously affect the estimation of the relevant parameters of interest. Once more, this involves questions of judgment rather than of mathematical statistics, and the price to pay is that the traditional global tests of goodness-of-fit lose importance and may become meaningless. Often, despite huge data sizes, preciously few degrees of freedom are relevant for the important parts of the model, while the majority, because of their sheer number, will highlight mere model inadequacies in the less important parts. We may be not willing to spend the effort, or perhaps not even able, to model those irrelevancies. We need local assessments of the relevant parts of the fit, whether informally through plots of various kinds, or formally, through tests. Compare also the incisive remarks by Box (1980).

Parameter estimation in a composite model is tricky. Unfortunately, theoretical results applying to the components rarely carry over to the composition. Usually, some version of backfitting will perform part of the work: assume there are component models $A_1, ..., A_n$, and you know how to estimate the parameters of A_i separately for each i. Fix preliminary parameter estimates for all models except A_i, improve those of A_i, and cycle repeatedly through all i. This should result in correct values of the parameter estimates. However, backfitting creates devilishly tricky statistical problems with regard to the estimation of standard errors of the estimated parameters, and with counting the number of degrees of freedom for goodness-of-fit tests. The standard errors estimated from the separate models A_i, keeping the others parts of the model fixed, may be serious underestimates. Cross-validation estimates, for example, are invalidated by repeated cycling.

Once I had felt that stochastic modeling, despite its importance, belonged so much to a particular field of application that it was difficult to discuss it in a broad and general framework, and I had therefore excluded it from a discussion of current issues in statistics (Huber 1975a, p.86). I now would modify my former stance. I still believe that an abstract and general discussion will fail because it is practically impossible to establish a common basis of understanding between the partners of such a discussion. On the other hand, a discussion based on, and exemplified by, substantial and specific applications will be fruitful. All my better examples are concerned with the modeling of various applied stochastic processes. This is not an accident: stochastic processes are creating the most involved modeling problems. The following example (on modeling the rotation of the earth) may be the best I have met so far. It shows the intricacies of stochastic models in real situations, in particular how much modeling and data processing sometimes has to be done prior to any statistics, and how different model components must and can be separated despite interactions. The runner-up is in biology rather than in geophysics, and it operates in the time domain rather than in the frequency domain (modeling of circadian cycles, Brown 1988).

5.7   MODELING THE LENGTH OF DAY

This case study, excerpted from Huber (2006), illustrates not only the complexity of some models, but also the interplay of inspection, modeling, simulation, comparison, model fitting, parameter estimation and interpretation. I believe the main interest of this example lies in the fact that – thanks to the complexity of the model! – its inaccuracies could not only be identified, but also kept under control.

The starting point of the investigation had been a figure in a paper by Stephenson and Morrison (1995) on the length of day (LOD), here reproduced as Exhibit 5.1. The underlying physical problem is that because of tidal friction, the rotation of the earth slows down: the length of day increases by about 2 milliseconds per century. The analysis by Stephenson and Morrison of medieval and ancient eclipses back to about 700 BC had shown that on top of the systematic slow-down there are very substantial random fluctuations. They must be taken into account when one is extrapolating astronomical calculations to a more distant historical past. In particular, one is interested in the cumulative version of the LOD, namely the difference $\Delta T = ET - UT$, where ET, or ephemeris time, is the uniform time on which astronomical calculations are based, while UT, or universal time, corresponds to Greenwich mean time and is based on the irregular rotation of the earth. ΔT increases quadratically with time and by 2000 BC reaches approximately 13 hours – note that this interchanges day and night!

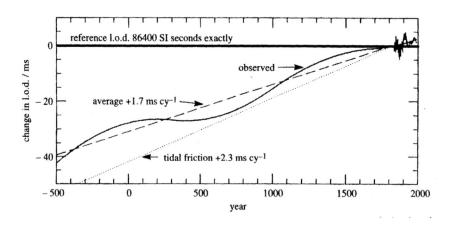

Exhibit 5.1 The changes in the LOD from −500 to +1990 were obtained by taking the first time derivative of a spline curve fitted to the observed values of ΔT. The high-frequency changes in the LOD +1830 to +1990 are taken from Jordi et al. (1994). Note the waves in the LOD with a wavelength of about 1500 years and an amplitude of about 4 ms. Reproduced with permission, from Stephenson and Morrison (1995, p. 197, Fig. 7).

Exhibit 5.1 reminded me of the Slutsky (1927) effect – namely that smoothing can create spurious cycles – and of some in-class demonstrations I had made with heavily smoothed first-order auto-regressive AR(1) processes. Therefore, I wondered whether such a process

$$X_t = \alpha X_{t-1} + \sigma Z_t \qquad (5.2)$$

where the innovations Z_t are assumed to be independent normal random variables with mean zero and variance one, with t in steps of 1 year, might serve as a simple phenomenological model for the stochastic components of the LOD process. A few simulations with various choices for α and σ showed that this was indeed the case, with α very close to 1. Note that $\alpha = 1$ corresponds to a Brownian motion (or random walk) model. In terms of the LOD-process, these poorly determinable fluctuations thus seemed to be compatible with a Brownian motion model, whose increments could be estimated crudely from the figure to have a variance of about $0.05 \ ms^2/year$.

I wondered whether such a millennial Brownian motion component was a long-range effect only, or whether it would be discernible also in the more accurate but much shorter modern series of measurements, and in particular whether those modern measurements might even permit a more accurate estimate of the variance of the increments, and hence lead to a more accurate estimate of the extrapolation error of the LOD and of ΔT.

With regard to the "philosophical" aspects of data analysis, some background information on experiences I made during this not entirely straightforward but ultimately successful investigation may be of interest. It ranged, off-and-on, from 1996 to 2006, and I had to dig much deeper into the underlying subject matter than I would ever have anticipated. Over this time, I accessed some six modern data collections, most of them publicly accessible over the internet, of varying length and quality, with LOD values in 4-month, 5-day and 1-day intervals, respectively. The 4-month series started in 1830, the others in 1962 or later.

The somewhat naive assumption that a close look at the behavior of the modern LOD process data in the time domain would suffice to show the putative Brownian motion component could be rejected almost immediately. It soon became evident that while modeling and simulations could be done in the time domain, most of the analysis would have to happen in the frequency domain.

Fortunately, I had learned the high art of spectrum analysis early in my life (in connection with a research project on the spectrum and bispectrum analysis of electroencephalograms). I had learned to taper the data (after having been badly burnt by neglecting it) and to correct the biases caused by tapering, and I had realized that when one is choosing a smoothing window, the psychology of vision is just as important as statistical optimality: approximately Gaussian windows produce the least misleading visual cues.

Clearly, the historical data of Stephenson and Morrison were too crude, and the modern data segments too short, to permit discrimination between an AR(1) process with a relaxation time of, say, 500 years ($\alpha = 0.998$) and a Brownian motion ($\alpha = 1$). With regard to the ultimate goal of the investigation (estimating the extrapolation error), the latter would be a conservative choice: Brownian motion makes extrapolation hardest. There is some weak circumstantial evidence from plate tectonics, suggesting a relaxation time in the order of 4000-5000 years.

In the power spectrum of the LOD values, a Brownian motion component manifests itself as a difficult to analyze singularity at frequency 0. In the power spectrum of a differenced series of LOD values, it would manifest itself as an easier to analyze horizontal tail end in the low frequency part of the spectrum. But it soon turned out that a quantitative treatment of this putative Brownian motion component would not be easy, since there were decadal fluctuations with waves whose length was of the order of 40 years, see Exhibit 5.2. While these waves look similar to those of Exhibit 5.1, their amplitude is 2-3 times larger than what one would expect from self-similarity of a Brownian motion process, so the two processes appear to have different origins. The decadal waves would dominate the low frequency spectrum in the relatively short stretches covered by the modern data. Then there were strong seasonal effects at 1 year and at 6 months, and the so-called "50-day Madden-Julian oscillation" dominated the spectrum for periodicities below 8-10 months, see Exhibits 5.3 and 5.4. These graphs plot the logarithm of the estimated spectral ordinates (logarithms in order to better cover the wide power range, and to equalize the random variability) against the square root of the frequency (square roots in order to spread the interesting low frequency end of the spectrum). At best, a Brownian motion contribution would become visible in an intermediate periodicity range between the seasonal effects and the decadal waves, roughly between 1 and 10 years. But the other contributions might leak into that range, and it became clear that it would be necessary to model the other contributions and to correct for possible biases introduced by them.

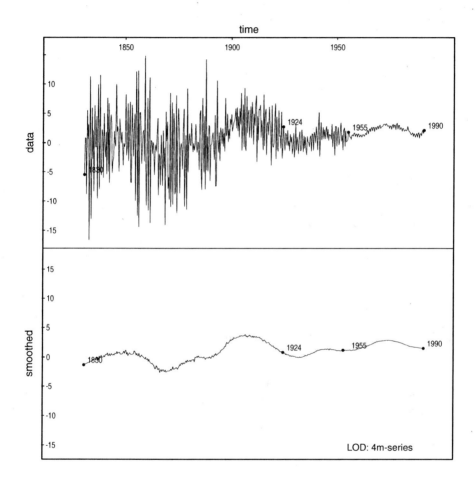

Exhibit 5.2   The 4-lunation series (covering the years 1830-1990 in 4-month intervals) in the time domain: the actual data in the series, and a smoothed version (obtained by forming moving averages spanning 31 values, i.e. 10 years). Shown is the difference of the length-of-day from 86400 seconds, in milliseconds. Note the changing level of the observational noise, the decadal waves and the seasonal effects (the latter visible after 1955). The drift of 1.7 ms/cy is barely perceptible.

Initially, my modeling had been entirely phenomenological. The decadal fluctuations, modeled from the 4-month series, and the 50-day oscillations, modeled from the 5-day series, produced broad spectral humps and could be represented satisfactorily by AR(2) processes. Later, my modeling became progressively more physical. I tried to construct putative physical models explaining the phenomena. While such models need not be "true" in an objective sense, they at least should be physically possible and obey basic physical laws, such as preservation of angular momentum, which the earlier phenomenological AR(2) models had neglected. The physical models – systems of linked, randomly excited, damped harmonic oscillators – led to systems of second order differential equations, whose solutions in turn could be closely approximated by AR(2) schemes. The latter were used for simulations, that is, for producing synthetic LOD processes.

There were preprocessing artifacts. Most of them, but not all, were irrelevant for my purposes. Some of them were discovered only by comparing results from different data sets. In the 5-day series, the spectrum had been depressed by filtering in the periodicity range just below 30 days; this biased the AR(2) parameter estimates for the 50-day oscillations and made the spectral hump a little narrower. In the 1-day series, an analogous effect was in operation but affected only periodicities in the irrelevant range below 10 days; compare Exhibits 5.3 and 5.4 and note the shift in the cross-over point between the AR(2) model and the random walk process.

Aided by hindsight, a reasonably comprehensive description of the components of the LOD-process goes as follows. See Huber (2006) for details.

(1) Systematic drift of about 1.7 ms/cy, most of it caused by tidal friction. The "true" rate cannot be estimated very accurately from the data because of the random Brownian motion (2) sitting on top.

(2) Brownian motion (or random walk process). Putative cause: random changes in the rotational moment of inertia of the earth's mantle, induced by plate tectonics (with a relaxation time of at least several hundred years).

(3) Decadal fluctuations, with a root-mean-square amplitude of about 1.3 ms, causing a broad spectral hump centered near 45 years. See Exhibits 5.2 and 5.5. The common opinion is that the decadal fluctuations have to do with damped oscillations on the mantle-core boundary. Putative cause: random changes in the moment of inertia of the core, with a relaxation time of the order of 7 years.

(4) Seasonal effects, with root-mean-square amplitude of about 0.35 ms. Apparently, they have to do with the exchange of angular momentum between the atmosphere and the solid earth, caused by seasonal temperature changes and winds. See Exhibits 5.3 and 5.4. These are sharp spectral spikes, affecting

no more than 3 or 4 periodogram ordinates (the broader seasonal peaks in Exhibits 5.3 and 5.4 reflect the shape of the smoothing window). They can be taken out cleanly by fitting a trigonometric polynomial to the LOD-process.

(5) "50-day Madden-Julian oscillation". This is a broad spectral hump near period-icities of 40-50 days, corresponding to damped oscillations with a root-mean-square amplitude of about 0.15 ms. See Exhibits 5.3 and 5.4. Apparently, it is due to exchange of angular momentum between the atmosphere and the solid earth, mediated by winds and putatively caused by random temperature changes in the atmosphere, changing the moment of inertia of the latter. To obtain a quantitatively correct spectral hump, the temperature changes and the winds would have to decay (by radiation and friction, respectively) with relaxation times of approximately one week. The underlying physical mechanism can be modeled quite accurately by an AR(2) process. Incidentally, the RMS amplitude 0.15 ms of the LOD in this model corresponds to a RMS variability of global wind velocities of merely about 1 m/sec.

(6) Measurement errors. Extremely inhomogeneous, about 4 ms in the 19th century, 0.1 ms in the 1960s, and 0.02 ms around 2000. In the 4-month series the measurement errors dominate the spectrum for periodicities below 8 years (see Exhibit 5.5, cross-over at 3024 days). In the more accurate 5-day and 1-day series they are negligible in the spectral range of interest to us (above 10 days). See Exhibits 5.2 and 5.4.

(7) Solid earth tides. These are reasonably well understood, deterministic effects. In the later parts of the data series made available to me, namely since 1982, most of them had been eliminated through preprocessing, but not before (remnants are some peaks in the 10-14 days range in Exhibits 5.3 and 5.4).

(8) Side effects of preprocessing: suppression of high frequency noise, including measurement errors, through filtering. In the 5-day series they depress the spectral power for periodicities below 30 days, leading to biased estimates of the parameters of the AR(2) process (5). Their existence was discovered only when analyzing a shorter, very accurate 1-day series showing analogous artifacts, but in the range below 10 days.

There is a delicate interplay between the components (2) and (3). We note that random changes in the rotational moment of inertia of the earth's mantle, as postulated in (2), by preservation of angular momentum cause wiggles in the rotation rate of the mantle. These wiggles excite damped oscillations on the mantle-core boundary, with a resonance in the decadal range. High-frequency components only wiggle the mantle, but in the low-frequency range, mantle and core move together as one solid body. Even though the exact coupling mechanisms are not known, the net effect is that the spectrum of the differenced LOD-series (ignoring the 50-day oscillation) will be flat both below and above the resonance frequency, with a 17% smaller value in the low frequency range (the size of the drop is determined by the known ratio between the moments of inertia of mantle and core), with a hump of poorly determined shape in between. Incidentally, the physical modeling by second order differential equations showed that the resonance oscillations excited by the wiggles of the earth's mantle are too small to explain the observed decadal oscillations, which therefore must have at least one other cause.

The feature of interest in the spectrum of Exhibits 5.3 and 5.4 is the putative Brownian motion (or random walk) component, which should manifest itself in a flat low-frequency spectrum, squeezed in between decadal fluctuations, seasonal effects, and the 50-day oscillations. It can be estimated only from a periodicity range where neither of these three components is dominant. Leaving some safety margin, this range extends from 1 to 10 years and covers about 40 periodogram values, that is fewer than the 63 values covered by the smoothing window used in Exhibits 5.3 and 5.4. To confirm this range, we first must model the decadal fluctuations and the 50-day oscillations, in order to see whether and how far they leak into the range between them.

The accurate 5-day and 1-day series are too short to estimate the decadal fluctuations. In Exhibits 5.3 and 5.4 their contributions are attenuated by smoothing and are barely visible. The 4-month LOD series with its 160 years on the other hand is long enough that it can be used to fit an approximate AR(2) model to the decadal fluctuations (if we compensate for the Brownian motion contribution by iterative backfitting), but is too noisy to show any structure for periodicities below about 8 years. See Exhibit 5.5; the cross-over points between the Brownian motion (random walk) and the AR(2) model for the decadal fluctuations and the MA(1) model for the measurement errors are not very accurately determined; they are near 14 years (5040 days) and 8 years (3024 days) respectively. AR(2) models for the 50-day oscillations can be found from the 5-day and 1-day series, using periodogram ordinates between periodicities of, say, 20 and 90 days (i.e. in a range excluding the worst solid earth tides and the seasonal effects); for the actual calculation we must remember some tricks from the robustness trade, in order to take care of remnant spikes from solid earth tides.. The cross-over point between this AR(2) model and the Brownian motion is near 10 months in the 1-day series.

By the way, for somebody reared on Box-Jenkins time-series analysis, it is quite an educational experience if he has to devise his own ARMA estimates based on a periodogram segment! For that, one must calculate the spectrum from tentative ARMA parameters, compare this spectrum with the periodogram computed from the data, but restricting attention to the relevant segment, and then fit the spectrum to the periodogram by adjusting the ARMA parameters iteratively through nonlinear weighted least squares.

Of course, for all these estimates one must use iterative backfitting. It goes without saying that in order to avoid processing artifacts caused by smoothing, the parameter estimates were based on the unsmoothed periodogram values themselves. In the end, a reasonably accurate estimate of the parameter of the Brownian motion could be obtained from 40 periodogram ordinates for periodicities between 1 and 10 years, based on the best available 1-day series, which ranged from 1962 to 2006. This yielded a final estimate of $0.068\ ms^2/year$ for the variance of the increments of the Brownian motion, valid above the resonance frequency. For the millennial range (below the resonance frequency), this value must be reduced by 17%, which translates into $\sigma^2 = 0.056\ ms^2/year$, with a 95% confidence interval (0.036, 0.088). This is remarkably close to the crude value $0.05\ ms^2/year$ estimated from Exhibit 5.1, lending credence to my initial conjectural suggestion. The estimate of σ then has a standard error of about 12%.

The final results now can be summarized as follows. In the range above the decadal fluctuations, extending from about one hundred to several thousand years, the LOD is fairly well modeled by a Brownian motion with drift; the increments have an estimated variance $0.056\ ms^2/year$. An inspection of the behavior of the process in the time domain shows that the underlying *physical* Brownian motion (where independence and normality of the increments break down for short time intervals) would have to be driven by pulses distanced by at most a few years. The cumulative process, that is the clocktime correction ΔT, then is a Gaussian process with known covariance structure. If ΔT is known at isolated points in time thanks to historical eclipse observations, this covariance structure now permits to interpolate and extrapolate ΔT by calculating conditional expectations, and to estimate the associated errors. Even though this model is only approximate, it puts interpolation and extrapolation on a much firmer basis than the hitherto used spline interpolation. Thanks to eclipse observations, ΔT is now known to within a standard error of about 5 minutes back to 700 BC, see Huber and De Meis (2004). By 2000 BC, the drift of ΔT extrapolates to about 13 hours. The Brownian motion model makes it possible to attach an estimated standard error of about 1 hour ($\pm 12\%$) to these 13 hours. By the way, if we assume a relaxation time of 500 years, the extrapolation error is reduced from 1 hour to about 23 minutes. This could be put to good use: it helped to identify a solar eclipse observation from the time of Sargon of Akkad (June 14, 2353 BC), see Huber (2011).

Among many other things, this example illustrates that for some parts of the model (usually the less interesting ones) we may have an abundance of degrees of freedom, and a scarcity for the interesting parts.

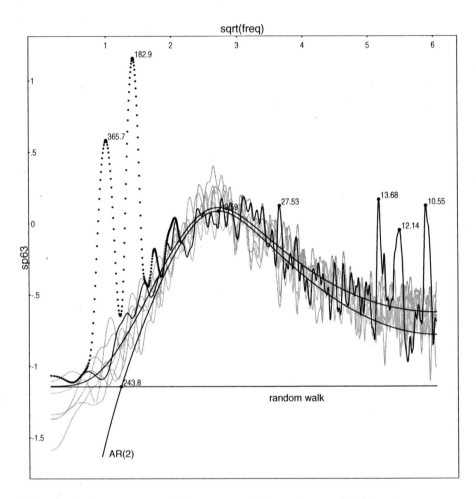

Exhibit 5.3 Log10 spectrum of the differenced 5-day series (covering the years 1962–1995 in 5-day intervals). The cross-over between the random walk process and the AR(2) model occurs near 8 months (243.8 days). All spectra are smoothed with a Gaussian window with 63 coefficients (sp63). Ordinate units: Log10 of $ms^2/year$, from $10^{-1.5}$ to $10^{+1} ms^2/year$. Shown are the spectra of: the raw data (dotted, the dots correspond to the abscissae of the periodogram values), the de-seasoned data (solid, with a few selected periods indicated in days), six simulations of the model (gray), the AR(1) random walk model, with $\sigma^2 = 0.072 ms^2/year$, the AR(2) "50-day" oscillation; the model is the superposition of the latter two processes (on purpose, only the two most prominent model components (2) and (5) were used). (With kind permission from Springer Science+Business Media: Huber (2006, p. 292, Fig. 9).)

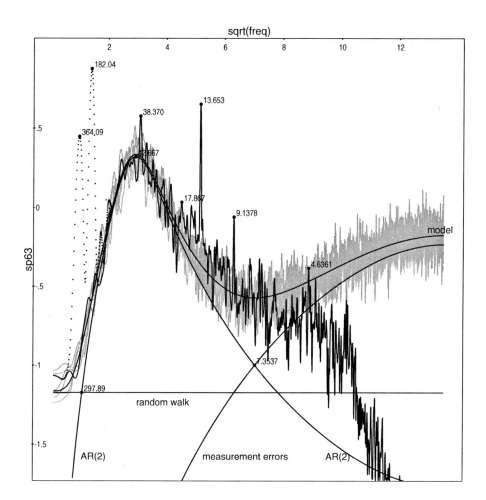

Exhibit 5.4 Log10 spectrum of the differenced 1-day series IERS-C04 (January 1962–March 2006), smoothed with 63 coefficients (sp63). The cross-over between the random walk process and the AR(2) model occurs near 10 months (297.89 days). Ordinate units: Log10 of $ms^2/year$. Shown are the spectra of: the raw data (dotted, the dots correspond to the abscissae of the periodogram values), the de-seasoned data (solid, with a few selected periods indicated in days), six simulations of the model (gray), the AR(1) random walk model, with $\sigma^2 = 0.067ms^2/year$, the AR(2) "50-day" oscillation, MA(1) measurement errors; the model is the superposition of the latter three processes. Note the suppression of the noise for periodicities shorter that 10 days. (With kind permission from Springer Science+Business Media: Huber (2006, p. 294, Fig. 11).)

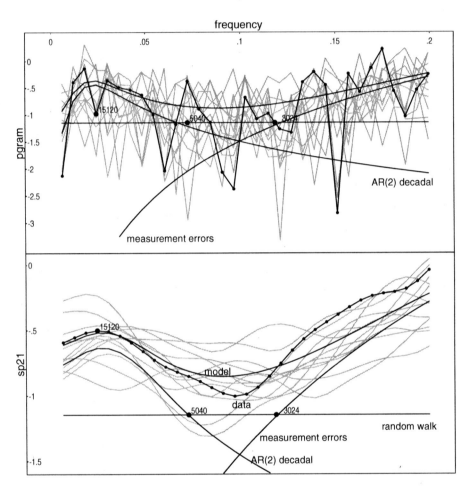

Exhibit 5.5   Low-frequency part of the Log10 periodogram (*upper panel*) and Log10 spectrum (*lower panel*, smoothed with 21 coefficients) of the differenced four-lunation series. Ordinate units: Log10 of $ms^2/year$. Note that these graphs use a frequency rather than a root-frequency scale (unit: $year^{-1}$). Shown are the spectra of: the raw data (*dotted*, the dots correspond to the abscissae of the periodogram values), 12 simulations of the model (*gray*), the AR(1) random walk model, with $\sigma^2 = 0.072ms^2/year$, the AR(2) decadal fluctuations, the MA(1) measurement errors, the model = superposition of the latter three processes. (With kind permission from Springer Science+Business Media: Huber (2006, p. 291, Fig. 8).)

5.8   THE ROLE OF SIMULATION

Following Karl Pearson's classical example, goodness of fit usually is assessed in terms of weighted sums of squared deviations of the observations from their expected values; their distribution then is approximated by a χ^2 distribution with a suitable number of degrees of freedom. As already Pearson had found, the distribution theory of this test statistic is not exactly trivial, and with complex, composite models, this kind of comparison is no longer feasible for multiple reasons. The fact that iterative backfitting defies distribution theory (except in trivial cases) is only one of them. Another serious problem is due to artifacts caused by preprocessing and processing of the data. Remember that a statistician rarely sees the raw data themselves – most large data collections in the sciences are being heavily preprocessed already in the collection stage, and the scientists not only tend to forget to mention it, but sometimes they also forget exactly what they had done.

A possible, and often the only escape from this quandary is simulation: compare the data with simulations of the model. (Note that in the goodness-of-fit context resampling methods are inappropriate, quite apart from the fact that they do not work for highly structured data.) This way, one can derive a test from any arbitrary statistic: calculate the statistic and adjust nuisance parameters in exactly the same way for the data, and do this for a large number of simulations of the model. Reject if the value of the statistic derived from the data is sufficiently far out in the tails of the set of values derived from the simulations. In particular, the spread of the simulated values of a point estimate permits to supplement the latter with an estimated standard error. Of course, just like the classical approaches, also the simulation methods only measure stochastic variabilities intrinsic to a model assumed to be correct: they implicitly assume that the estimated parameter values are so close to the true ones that the latter can be replaced by the former without committing a serious error when assessing the variability of an estimate or test statistic. Admittedly, in practice there may be problems; for example, parameter estimation may fail to converge for a small percentage of the samples, and this may selectively affect the tails.

Under mild monotonicity assumptions, but at a relatively high computing cost, it is possible to supplement point estimates with somewhat more reliable confidence interval estimates than by the method just described. For example, in order to find a lower confidence bound, one takes the model and replaces the estimated value θ of the model parameter of interest (in the example of the preceding section: the variance of the increments of the Brownian motion) by a suitably chosen smaller value θ'. Then one uses the thus modified model for simulating 1000 data sets, and derives new parameter estimates from each of these sets. If, say, 950 of the newly estimated values of θ then are smaller, and 50 larger than the original estimate of θ, then θ' constitutes an approximate one-sided 95% lower confidence bound. The determination of a suitable θ' is expensive because it requires a considerable amount of trial and error.

But thanks to simulation, a judgmental assessment of goodness-of-fit need not even be based on a test statistic (whose selection always is delicate). The principle is simple and inspired by the line-up methods used by the police: if the actual data hide well among half a dozen or a dozen simulations of the model, the model is judged acceptable; if the actual data stick out like a sore thumb, the model is no good. Exhibits 5.3 and 5.4 illustrate the approach by showing the spectrum estimates resulting from 6 simulations of the model, indicating a good fit in the interesting low frequency range. But the example also illustrates a general problem of any global approaches to goodness-of-fit, namely the peaks in the 10-14 day range, which make the actual data set stick out like the sore thumb mentioned above. In this case the origin of the discrepancy is understood, and it is irrelevant because it lies in an uninteresting frequency range. — A more sophisticated version of the line-up idea, permitting approximate significance tests, is to create a pool composed of the actual data set and, say, 99 simulated sets. Somebody not knowing which is which has 5 attempts to pick the actual set out of the pool, using any tools of his or her choice. If the actual data set is not among the selected five, the model is deemed to be adequate.

5.9   SUMMARY OF CONCLUSIONS

If we want to be able to deal with increasingly larger and more complex data sets, we need to go beyond the current, overly narrow ingrained modeling concepts of statistics. The models will become more complex, too, but a clean statistical theory with mathematically rigorous results is possible only for clean and simple models. We will have to pay a price, but we will also gain something in the process. The questions will shift more and more from a mere yes-or-no global check whether the model is adequate (I prefer the term "model adequacy" to "model validity" – a model adequately rendering the observations need not be valid in any intrinsic sense), to a detailed assessment of the quality of the fit, to questions of interpreting the fit, and in particular to the need to locate and interpret deviations from a model which is known to be imprecise, and to separate essential deviations from irrelevant ones.

CHAPTER 6

PITFALLS

We already have encountered some pitfalls of data analysis. In one example a problem arose because one had overlooked that carriers and controls had not been properly matched (see Section 2.3). Another example was concerned with the fact that it is no longer possible to calculate reliable P-values *after* one has looked at the data (see Section 2.4.4). This does not mean that one should not look at the data – on the contrary – but merely that one should not fool oneself and others by spurious P-values. The present chapter offers examples of three further common pitfalls: Simpson's paradox, problems caused by unrecognized missingness, and conceptual pitfalls of regression. For the sake of clarity, they shall be illustrated with the help of relatively small examples. But with larger data sets all these pitfalls become more difficult to recognize, and I believe that with the advent of data mining, that is, with the unsupervised grinding through massive data sets, they have become particularly pernicious. Good traps are camouflaged, and data analytic pitfalls are no exception – they often hide behind the smokescreen provided by a complex application.

Data Analysis: What Can Be Learned From the Past 50 Years. By Peter J. Huber
Copyright © 2011 John Wiley & Sons, Inc.

The problem with Simpson's paradox is that only few textbooks and hardly any statistics courses draw attention to it.

Missing values often are considered as a mere nuisance because a data matrix with holes in it cannot be handled by the standard formulas of linear algebra. Therefore, some (hopefully sensible) values are filled into those holes by imputation. Typically, one does so by calculating conditional expectations, given the non-missing values, often assisted by iterative backfitting. The so-called EM algorithm is an example for such a procedure. But really awkward problems arise if one is not aware that the data set is missing certain values.

Regression is one of the oldest data analytic tools (under the label "Method of Least Squares") and one of the most powerful. Like all powerful tools it is prone to be misused. One such misuse was mentioned in Section 2.4.2: use of regression instead of survival analysis. The main problem with regression is that many regard it as *the* method for measuring and describing the dependency of a variable on others, which it is not.

6.1 SIMPSON'S PARADOX

Already in 1940, Deming had admonished the statistical profession that as a whole it was paying much too little attention to the need for dealing with heterogeneous data. This still holds true, despite Simpson's seminal paper (1951), and the problem has been aggravated by the advent of data mining and of blind, automated procedures, which are almost guaranteed to run into Simpson's paradox without the statistician ever knowing. Two simple examples with synthetic data show how drastic the effects can be.

The first synthetic example concerns expenditures on cosmetics versus personal income. Exhibit 6.1 shows the data and three regression lines. Women tend to spend more on cosmetics, even though their income tends to be lower. If we group men and women together and regress expenditures for cosmetics against income, we may find a decreasing relationship: if you earn more, you seem to spend less!

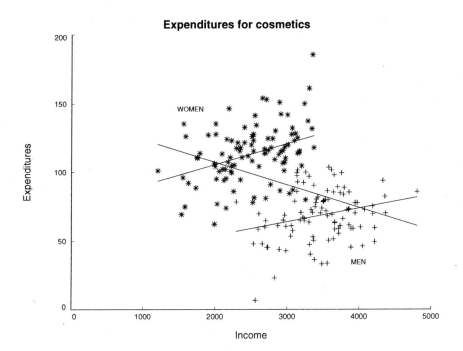

Exhibit 6.1   Simpson's paradox: Expenditures on cosmetics versus income (synthetic data). Women (*), Men (+).

The second example is about apparent sex bias. Exhibit 6.2 uses fictitious data to show an extreme case of apparent sex discrimination. Note that the aggregate of the two departments seems to show a startling bias in favor of men: the admission rates are 68% for men and 36% for women, but separately, the departments show a bias in favor of women.

Economics	Men	Women
admitted	99 (77%)	6 (86%)
rejected	30 (23%)	1 (14%)

Biology	Men	Women
admitted	3 (14%)	44 (33%)
rejected	18 (86%)	88 (67%)

Total	Men	Women
admitted	102 (68%)	50 (36%)
rejected	48 (32%)	89 (64%)

Exhibit 6.2   Admission rates (synthetic data).

An excellent case study with real data is presented by Freedman et al. (1991, Chapter 2). At the University of California in Berkeley, the associate dean of the graduate division had noticed an apparent bias against women in graduate admissions: the aggregate admission rates had been 44% for men and 35% for women. The issue then had been cleared up with the help of two colleagues from the statistics department, see Bickel, Hammel and O'Connell (1975). In this case, the six departments, when taken separately, showed no clear pattern. Major by major, there did not seem to be any bias against women. Some majors favored men, but others favored women. On the whole, if there was any bias it ran against the men. Apparently, there had been a self-selection effect: the men were preferentially applying to easy majors, while the women had preferentially applied to departments where it was hard to get in.

6.2 MISSING DATA

Large data sets rarely are complete. The reasons why some data are missing vary. Maybe there is a question of funding. Maybe in the course of a study some variables have been dropped from, or added to (say for political reasons, e.g. questions of race). Maybe part of a data archive has been destroyed. Maybe some people have refused to answer certain questions of a questionnaire. Maybe the values to be measured fall outside of the range of the measuring device. Maybe obduction results are not available because the patient is still alive. Subtle but pervasive cases of missingness occur when there are systematic differences between the population one would like to observe and the smaller subset which is accessible to observation.

If presented in isolation, missing value problems usually are easy to spot, even if they are not easy to correct. However, in the context of an actual analysis of real data, missingness often is not recognized as such because the missing values fall outside of the event horizon of the observer and do not exist in his or her frame of mind.

Ready-made recipes for dealing with missing data are available only in a few exceptional cases. Standard statistics texts rarely mention the problems caused by missing data. Therefore, the problems connected with missingness often are ignored and swept under the carpet. Little and Rubin (1987) is the principal reference work on missing data.

The only general case that is under theoretical control occurs in regression-like situations. Assume that some values of the independent variable y are "missing at random" (MAR). This is supposed to mean: the event that a particular y is missing may depend on the independent variable x, but not on the unknown value of y which otherwise would have been observed. This is a colloquial definition leaving out the fine print.

Here is an attempt to spell out a formally precise version of MAR. Let us describe the situation by a collection of random variables $(X_i, Y_i, Z_i), i = 1, \ldots, n$, where $X_i \in \mathbb{R}^p$ and $Y_i \in \mathbb{R}$ formalize the "independent" and "dependent" variables respectively, and $Z_i = 1$ if Y_i is missing and 0 otherwise. We take MAR to mean that given $(X_1, \ldots, X_n)$ the random variables $Y_1, \ldots, Y_n, Z_1, \ldots, Z_n$ are conditionally independent. Since only conditional probabilities are involved, it does not matter whether the X_i are random variables or fixed quantities (as in a designed experiment). Note that we explicitly require that the "dependent" variables are conditionally independent; this goes beyond what is implied by the vague colloquial definition. Unfortunately, even a weak form of the latter is hardly ever satisfied in practice.

The most insidious type of missingness occurs if one is not even aware that the data set is missing some values. We encountered such a case in a very large data set on highway maintenance. It happened that sometimes one had forgotten to record an intervention (repair), an omission which left absolutely no trace in the data set, except perhaps in the form of a longer than usual interval between repairs.

The following two case studies are interesting because in the first case the original investigator inadvertently had fallen into the trap, but the missing values really were available elsewhere in the data set. In the second case, overlooked missingness had caused totally surprising qualitative effects.

6.2.1   The Case of the Babylonian Lunar Six

I first present a simple synthetic example that illustrates the salient issues in isolation. We take bivariate normal data (X, Y) with mean 0 and covariance matrix

$$\begin{pmatrix} 1 & 0.9 \\ 0.9 & 1 \end{pmatrix}$$

If we regress Y on X, we of course obtain a regression line with slope 0.9 and intercept 0. But now assume that only values $Y \geq 0$ make sense and are visible. Then, regression of Y on X gives a slope of 0.68 and an intercept of 0.31, see Exhibit 6.3. Only very few people confronted with such a data set will realize that they are missing part of the evidence.

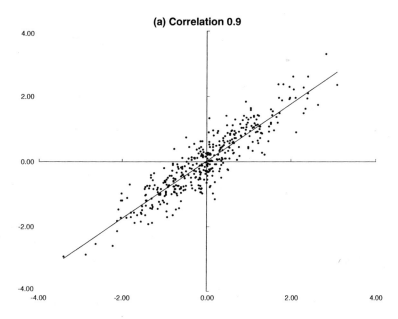

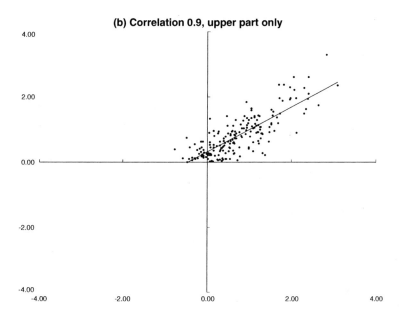

Exhibit 6.3   (a): all data; (b): visible data only ($Y \geq 0$).

The following is an example with real data, where this problem occurs in a complicated context, and where the first investigator inadvertently fell into the trap. The beauty of the example is that the "missing" data really are available.

The data sets in question consist of a large number of ancient timings of moonset and moonrise relative to sunset or sunrise. The numbers are recorded on Babylonian clay tablets from the last few centuries BC. For the general background see the articles in *The Place of Astronomy in the Ancient World*, ed. by D. G. Kendall (1974), in particular those by Sachs (p.43-50) and Stephenson (p.118-121). The observational data are called "Lunar Six", because six different phenomena were observed, traditionally designated by their ancient terms NA, SHU2, NA, ME, GE6, KUR (the numbers differentiate between cuneiform signs sharing the same pronunciation; in the graphs the discriminating numbers were sometimes omitted). Perhaps it would be more convenient for the reader if we had used mnemonics: MS-SS instead of NA for denoting the time difference MoonSet minus SunSet, etc., but I use these mnemonics reluctantly, they give away the solution of the puzzle:

(1) NA = MS-SS, time from sunset to moonset (after new moon, on the first evening when the lunar crescent was visible).

(2) SHU2 = SR-MS, time from moonset to sunrise (near full moon, on the last morning when the moon set before sunrise).

(3) NA = MS-SR, time from sunrise to moonset (near full moon, on the first morning when the moon set after sunrise).

(4) ME = SS-MR, time from moonrise to sunset (near full moon, on the last evening when the moon rose before sunset).

(5) GE6 = MR-SS, time from sunset to moonrise (near full moon, on the first evening when the moon rose after sunset).

(6) KUR = SR-MR, time from moonrise to sunrise (before new moon, on the last morning when the moon was visible).

It is generally believed that the measurements were made with some sort of water clock, but we do not know in detail how – perhaps by collecting the water dripping out of a container and then weighing it, or perhaps by measuring water levels, either in the upper or lower container. There are many potential sources of bias, for example the flow speed would depend on the water level.

The times are given in time degrees: $1° = 4$ minutes. Note that in 1 time degree the celestial sphere rotates 1 degree of arc. If the observations are plotted against the calculations, they should ideally lie on straight lines passing through the origin, with slope 1. Higher or lower slopes mean that the clocks are running fast or slow; deviations from a straight line indicate nonlinear behavior of the clocks, e.g. by gradually slowing down after refilling at a fixed time (e.g. sunrise).

For sound methodological reasons, the first investigator, the geophysicist F. R. Stephenson (1974), on the basis of a relatively small data set, had investigated the six different event types separately. But because of this, he inadvertently had fallen into the missingness trap. Such investigations are far from trivial, also the modern computations are afflicted by various uncertainties. He found puzzling, counter-intuitive patterns of the slopes and intercepts. He tried to explain them by assuming that the Babylonians used different collecting vessels for the different types of intervals.

Later on, at Stephenson's request, the Assyriologist A. Sachs excerpted an order of magnitude larger set of Lunar Six data from Late Babylonian astronomical diaries and other texts, a total of 724 values. They range from 301 BC to 78 BC and are disjoint from the data analyzed by Stephenson in 1974. The new data at first sight seemed to confirm his curious findings. But on closer scrutiny, it turned out that the features which had puzzled him – namely that the regression lines calculated by him for the six phenomena showed counter-intuitive intercepts and did not match up in the expected fashion – to a large part were due to the missing values effect mentioned before. In addition, there are also some problems caused by systematic errors affecting large time intervals. In the following I shall consider only the short time intervals (2) to (5) near full moon and disregard the longer ones (1) and (6) occurring near new moon.

The beauty of this data set is that the "missing" values actually are available! For example, SHU2 = SR-MS and NA = MS-SR both are concerned with the time difference from sunrise to moonset in the morning: if this difference is negative, its absolute value is called SHU2, if it is positive, NA. In other words, SHU2 and NA correspond to the lower and upper halves of the synthetic example, and similarly ME and GE6. In the following pictures, Exhibit 6.4 (a) and (b) show the data for SHU2 and NA, while Exhibit 6.5 (c) shows how they fit together, if SHU2 is given the negative sign. Exhibit 6.5 (d) shows the central part of (c) and how well the two join. The lines in figures (a)-(d) correspond to the ideally expected calculated values. The final Exhibit 6.6 shows the differences between observed and calculated values. There is heteroskedasticity (larger errors for larger time differences), and effects of nonlinearity also are clearly visible. The systematic differences of about $0.75° = 3$ minutes between observation and calculation probably imply that the observer was located much higher up than we had assumed (about 108 m above ground, instead of 20 m). But the scatterplots of SHU2 and NA, and of ME and GE6, repectively, join seamlessly, doing away with Stephenson's puzzle.

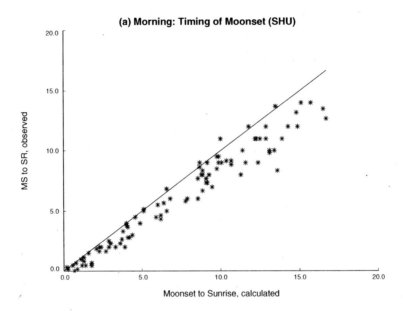

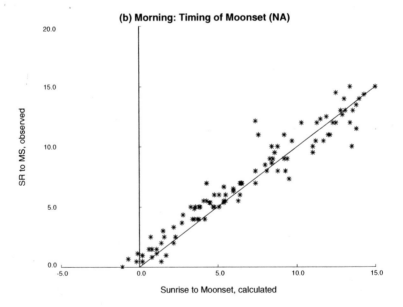

Exhibit 6.4   Lunar Six. Morning. (a): after sunrise. (b): before sunrise.

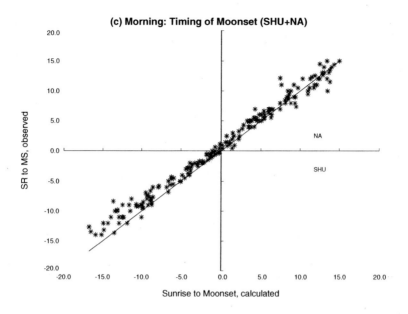

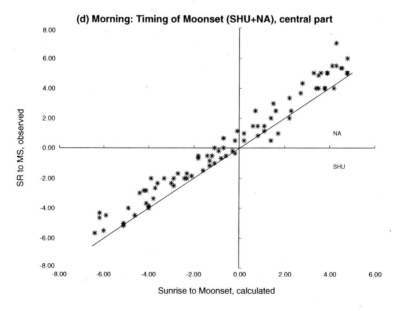

Exhibit 6.5   Lunar Six. Morning. (c): SHU2 and NA joined. (d): same, central part.

This finishes the case of the overlooked missing values. But we continue the analysis since – like the proverbial Princes of Serendip – we shall be able to pick up some interesting and scientifically relevant bits on the wayside.

By accident, the ancient Babylonian measurements derive from a remarkably well designed experiment, which permits a much better separation of systematic and random errors than what ordinarily is possible in opportunistic data sets. Results on the accuracy of ancient measurements are of interest for the history of applied science, and are a precondition before one can use timed Babylonian observations in a quantitative fashion (e.g. to check on current astronomical theories). This was my motivation for undertaking the Lunar Six data analysis reported in Huber (2000b). One such result is that the standard deviation of a single Lunar Six measurement t (expressed in time degrees) is well approximated by

$$\sigma = \sqrt{0.6^2 + (0.08\,t)^2},$$

that is, the error is about 0.6° or 2.5 minutes for short intervals, and about 8% for large intervals. The random variability of the short intervals must be due more to fluctuations in the atmospheric refraction than to timing errors. Note in particular that for the same calculated small time difference, observed differences of both signs do occur. From the Lunar Six measurements I estimated a RMS variability of the order of 0.20° in horizontal refraction, and I learned only afterwards that modern observers had found a value of 0.16° for the same effect (Huber 2000b, p. 234). Another result was that short intervals (below 6° or 24 minutes) were measured without perceptible systematic errors: the slopes in this range are not significantly different from 1. SHU2+NA give a slope of 0.988, and ME+GE6 give 0.995, with an estimated standard error of 0.020.

On average, relative to sun events, moon events occur about 10 seconds earlier than calculated. While this small difference is not significantly different from 0, it permits a welcome independent confirmation of the ΔT determined from lunar eclipses, see Section 5.7 and Huber and De Meis (2004, p. 24-28). The standard error of the determination of ΔT for 200 BC from the Lunar Six is about 5.7 minutes, from the eclipses about 2.6 minutes, and the 68% (or 1σ) confidence intervals of the two independent determinations overlap.

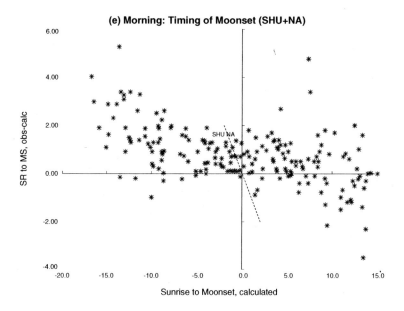

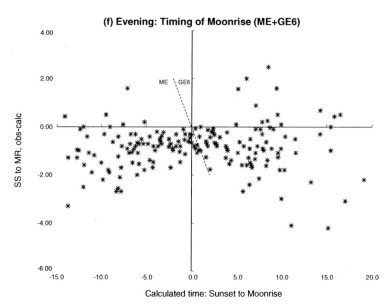

Exhibit 6.6   Differences observed–calculated. (e) Morning: Timing of Moonset (SHU2+NA). (f) Evening: Timing of Moonrise (ME+GE6).

6.2.2 X-ray crystallography

Missing data, when overlooked, usually cause quantitative effects in the form of unsuspected biases. The following is a weird example where missing data resulted in a totally surprising, massive qualitative effect.

A single crystal is a three-fold periodic structure; its electron density can be represented by a triple Fourier series

$$\rho(x, y, z) = \sum_{hkl} F_{hkl} \exp(2\pi i(hx + ky + lz))$$

where h, k, l are integers and $0 \le x, y, z < 1$ are coordinates in a (generally oblique) coordinate system adapted to the periodicities of the crystal. Such a crystal diffracts X-ray beams only in discrete lattice directions indexed by h, k, l, with intensities proportional to $|F_{hkl}|^2$. Thus, the absolute values of the structure factors F_{hkl} are observable, but not the phases. In the centrosymmetric case, the F_{hkl} are real, and

$$\rho(x, y, z) = \sum_{hkl} F_{hkl} \cos(2\pi(hx + ky + lz))$$

with

$$F_{hkl} = \sum_{s} f_s(\theta) \cos(2\pi(hx_s + ky_s + lz_s))$$

Here, the sum is over all atoms in the unit cell, $f_s(\theta)$ depends on the type of atom s and on the diffraction angle θ. The latter depends on h, k, l, but not on the coordinates (x_s, y_s, z_s) of atom s.

If one of the atoms is much heavier than the others, its contributions to F_{hkl} are preponderant. The coordinates of a single heavy atom are easy to determine, and it then is possible to find the other atoms by computing the electron density $\rho(x, y, z)$ from the above formula for the centrosymmetric case by combining the observed absolute values of F_{hkl} with the signs

$$\text{sign} \cos(2\pi(hx_s + ky_s + lz_s))$$

calculated from the heavy atom alone.

In the particular experiment, the values of $|F_{hkl}|^2$ were obtained from counter measurements as the difference between (background + signal) and (background), and standard deviations were determined from the counter statistic (Poisson distribution). There was one heavy atom (Thorium). In the particular case, the crystal was very small and the signal correspondingly weak, so the estimated value of $|F_{hkl}|^2$ often was negative. Somewhat arbitrarily, it was decided to ignore not only negative, but also small positive values of $|F_{hkl}|^2$, unless they exceeded their own estimated standard deviation, estimated from the Poisson counter statistics. Very surprisingly, the resulting Fourier synthesis showed not only the positions of several light atoms (Carbon and Oxygen), but also a few prominent peaks at physically very implausible locations. Actually, the highest spurious peak exceeded all genuine peaks (apart from the heavy atom itself).

Ultimately, after some lengthy detective work, it was found that the cause of this curious behavior was a missing value problem! Weak values with negative random errors were preferentially excluded, weak values with positive random errors preferentially included, and the net effect of this was that a diffuse cloud of positive random errors was given the heavy atom sign and added into the Fourier synthesis; there were several thousand coefficients. Now, the Fourier expansion of $\mathrm{sign}(\cos(x))$ is

$$\mathrm{sign}(\cos(x)) = \frac{4}{\pi} \sum_{k=0}^{\infty} \frac{(-1)^k}{2k+1} \cos((2k+1)x)$$

hence one can expect artifacts at odd multiples of the heavy atom coordinates (modulo 1): holes at multiples 3, 7, 11, ..., peaks at 5, 9, 13, ... Indeed, an alternating hole-peak pattern could be followed up to multiples above 20, see Exhibit 6.7. This explanation could be confirmed by two simple experiments of the what-if variety. First, since we had overall relatively weak intensities (because of the small crystal) against a high background count, the bias must have particularly affected a large percentage of the weak reflexions, preferentially occurring at high diffraction angles. This could be verified by remeasuring a sample of 49 weak reflexions. Of these, 13 were twice accepted, 10 twice rejected, and 26 switched sides; this confirmed that inclusion/exclusion of weak reflexions was determined mostly by the sign of the random measurement error. Second, by applying a stricter rejection criterion, the influence of the cloud of positive random errors presumably would be reduced, and indeed, the hole-peak pattern almost completely vanished. What had made the detective work devilishly tricky was that the graphics program plotting the electron density had suppressed negative densities as physically impossible, so the tell-tale holes initially had escaped our attention. See Huber-Buser (1969).

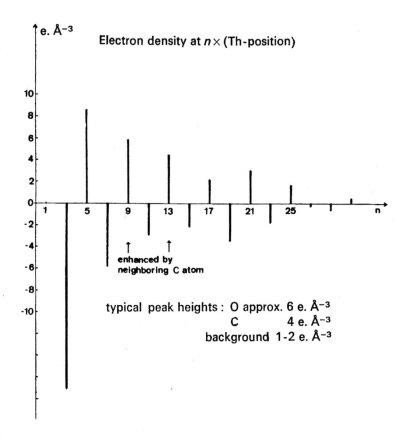

Exhibit 6.7   Hole-peak pattern. From Huber-Buser (1969).

6.3   REGRESSION OF Y ON X OR OF X ON Y?

If a question about the "correct" regression arises, that is, whether one should regress Y on X or X on Y, then this question most likely is ill-posed, and the answer is: neither! Some explanations are required.

By popular misapprehension, regression is a general method for fitting a straight line to X-Y-data. This makes sense if one of the variables, say X, is fixed and error-free – then one regresses Y on X. If both are subject to comparable random variability, then the method of choice is based on principal components – one fits by minimizing the sum of the squared orthogonal distances from the straight line. Of course, if the variability is small, then the difference between the different fits will also be small, perhaps too small to be physically relevant, but perhaps large enough to fool statistical significance statements. For an example where the choice does not matter, see Section 8.1, Exhibit 8.3.

An even more common misapprehension is that regression is *the* method for measuring and describing the dependency of Y on X. Regression indeed can be used for such purposes, but only in highly asymmetric cases, where all (or at least almost all) of the stochastic variability sits in Y, and X is fixed and (almost) error-free. Typical cases of this kind occur in designed experiments, where X is fixed by the design, or if X dcrives from a theoretical model (perhaps with with some free parameters, to be estimated) rather than from measurements.

The conceptual problem behind regression – of which few applied people seem to be aware – is that regression by its design is not a method for measuring the dependency of Y on X, where both are subject to randomness, but it should be described rather as a method that uses past observations of X and Y to find a good predictor, predicting the value of Y from a future observation of X. This prediction problem is a conditional one: predict Y, given X. In other words, we aim to predict the unknown Y by calculating the conditional expectation of Y, given X, and therefore, X can be treated as if it were fixed and error-free. With the traditional optimistic assumption that the observations are i.i.d. bivariate normal, we end up with linear regression as the optimal procedure.

The following non-trivial example from astronomy illustrates the issues. It is well known that the spectral lines in the light from far-away galaxies are shifted towards the red end of the spectrum. According to an almost generally accepted interpretation, this redshift is caused by a Doppler effect. That is, galaxies move away from us, with velocities v that increase with distance r. Whether or not this interpretation is correct, it is convenient to express the redshift in terms of the equivalent Doppler velocity v, and to measure it in km/sec. Unless we assume that our own galaxy occupies a special, distinguished position in the universe, it follows from this standard interpretation that the velocity must be proportional to the distance ("Hubble's Law"). But there is a highly controversial alternative theory, championed by the late I. E. Segal, who claimed that the redshift is not a Doppler effect. According to his theory the apparent v, as measured by the redshift, ought to increase quadratically with distance.

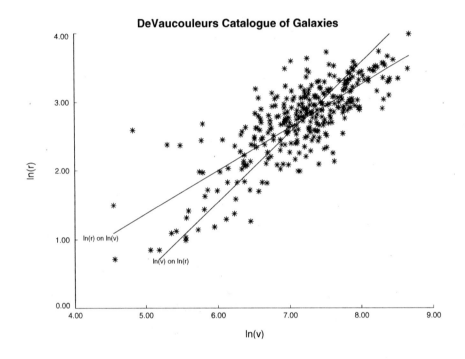

Exhibit 6.8   Regression of $\ln(r)$ on $\ln(v)$ versus regression of $\ln(v)$ on $\ln(r)$, where r is the estimated radial distance in megaparsec, and v is the estimated radial velocity in km/sec. Data from G. de Vaucouleurs (1979).

At first blush, it would seem straightforward to distinguish between the two theories on the basis of observational data, by checking the relationship between the logarithms of the redshifts and the logarithms of the radial distances. Exhibit 6.8 presents a corresponding scatterplot of galaxy data taken from G. de Vaucouleurs (1979).

With both theories the dependence of $\ln(v)$ on $\ln(r)$ is linear, according to Hubble's law with slope 1, according to Segal's with slope 2. But should one determine the slope in question from a regression of $\ln(v)$ on $\ln(r)$? Or should we regress $\ln(r)$ on $\ln(v)$?

Radial distances are difficult to determine and subject to various errors. They are estimated as $r = 10^{(mod/5-5)}$ megaparsec, where mod is the difference between the (estimated) absolute and the (measured) apparent magnitude. The apparent magnitude may be biased downward through intergalactic absorption, and the absolute magnitude is estimated on the assumption that certain objects have the same average magnitude and size throughout the universe. The farther away a galaxy is, the more difficult it is to identify such marker objects. There are missing values: the fainter a galaxy is, the more likely it is not observed; clearly, such observations are not missing at random.

On the other hand, redshifts (and the Doppler velocities v derived from them, expressed in km/sec) for all practical purposes are measured without error, that is within a margin of perhaps 1%, apart from occasional gross errors. A naive data analyst – recognizing only observational errors as sources of randomness and ignoring all other intrinsic sources – might therefore jump to the conclusion that he ought to regress $\ln(r)$ on $\ln(v)$, since the latter measurements are practically error-free. However, galaxies have substantial proper motions, evidenced by the existence of occasional blue-shifts, and there are clusters of galaxies moving more or less together; also our galaxy is part of a local cluster. Inside such clusters, the galaxies typically seem to move relative to each other with velocities of the order of 300-600 km/sec, and the proper motions of the clusters themselves, which are difficult to ascertain, may be even larger. It is by no means evident whether more of the variability sits in v or in r. If the former is dominating, we ought to regress $\ln(v)$ on $\ln(r)$.

If we take one of the best available, reasonably homogeneous data sets (G. de Vaucouleurs (1979)), comprising 322 galaxies, with v ranging from -69 to +5826 km/sec and r from 2 to 54 megaparsecs, and use ordinary least squares regression, either $\ln(v)$ on $\ln(r)$, or $\ln(r)$ on $\ln(v)$, we obtain the following regression equations (the $\pm$ values in parentheses are estimated standard errors):

$$\ln(v) = \quad 4.49(\pm 0.12) + 0.98(\pm 0.04) * \ln(r) + \text{noise},$$
$$\ln(r) = -1.78(\pm 0.20) + 0.63(\pm 0.03) * \ln(v) + \text{noise}.$$

The error estimates may be over-optimistic for many reasons. The first fit suggests the Hubble law, the second is closer to the alternative theory (where v is proportional to r^2)! See Exhibit 6.8.

The conclusion (at least: my conclusion) is that the de Vaucouleurs data do not suffice to discriminate between the standard Hubble theory and Segal's theory.

CHAPTER 7

CREATE ORDER IN DATA

> *Too little attention is given to the need for statistical control, or*
> *to put it more pertinently, since statistical control (randomness)*
> *is so rarely found, too little attention is given to the interpreta-*
> *tion of data that arise from conditions not in statistical control.*
> (W. E. Deming, 1940)

This chapter is concerned with techniques for creating order in data. Such techniques are helpful as first steps when one is confronted with the interpretation of inhomogeneous data, or, to use Deming's words, with the "interpretation of data that arise from conditions not in statistical control".

The methods I shall describe are not geared towards giving quantitative statistical results (such as confidence levels), but rather towards providing qualitative intuitive insights with the help of interpretable graphical layouts. The figures reproduced here

Data Analysis: What Can Be Learned From the Past 50 Years. By Peter J. Huber
Copyright © 2011 John Wiley & Sons, Inc.

were chosen to illustrate such layouts. Most of them (those with square frames) belong to the category of exploration graphs (see Section 2.6.2) and were produced by a Postscript hardcopy facility.

7.1   GENERAL CONSIDERATIONS

Assume you have a large pile of data without obvious structure. Most likely it will be heterogeneous. The standard first data analytic step, an unaided preliminary inspection of the raw data, almost inevitably will be confusing. Prior to any analysis and interpretation we must create some order. The common underlying idea is to arrange the data in such a way that items that are "similar" in some sense become neighbors. Interpretation is made easier if the order is represented graphically as a scatterplot, such that similar items are mapped into neighboring dots.

The resulting arrangement may (1) take the form of a linear ordering, or (2) of a 2-dimensional layout as in a topographical map, or (3) perhaps of a 3-dimensional layout if you have the necessary graphical facilities for dealing with 3-d, or (4) the data points are separated into disjoint clusters, or (5) the data are presented in the form of a tree. The main approaches to achieve such groupings or orderings are as follows.

Clustering techniques either attempt to separate the data into disjoint clusters, or to reconstruct an underlying tree structure. In my experience data only rarely can be separated into disjoint clusters, for example I learned from a study of human growth that anthropometric measurements do not separate boys and girls – the two clusters overlap. Hierarchical clustering works well if the there is a true underlying tree structure, for example in biology, if the task is to reconstruct an evolutionary tree. The results of clustering do not easily lend themselves to intuitive graphical representations. In the absence of an underlying tree structure, hierarchical clustering may impose an artificial structure that creates more confusion than clarity of insight. Very often there is a continuous graduation in the data, from small to large, or from conservative to progressive, or the like. Splitting such graduations into groups may be convenient, but is arbitrary. A computer generated continuous arrangement, leaving the splitting to the data analyst, may be preferable.

Principal component methods typically are based on the singular value decomposition. They will be discussed in Sections 7.2 and 7.4. They are not designed to group similar items together, but rather to find informative linear projections of a data set into a lower dimensional space. But they often achieve good groupings. The idea behind them is that, hopefully, systematic features present in the data will cause the widest spread, while random effects will cause lesser spread (and will be mapped into the suppressed orthogonal complement of the image space). Principal component methods work both with continuous and categorical data.

Projection pursuit methods try to find "interesting" low dimensional linear projections of a data set by numerically optimizing a certain objective function or "projection index". Projection pursuit comprises principal component methods if one takes sample variance as the projection index, but usually one restricts attention to scale invariant indices. Typically one will search for least normal projections, see Huber (1985b) for a survey and for the general theory. If we take them in this narrower sense, projection pursuit methods are suitable only for continuous data, and unfortunately, in high dimensions their computational complexity explodes. But they can be put to good use in low-dimensional cases if the principal component approaches fail, for example if the leading principal components happen to be equal or nearly equal.

Multidimensional scaling methods, to be discussed in Section 7.3, are closest to the original idea of grouping "similar" items together. Rather than using coordinate information, they use "similarity" of the items directly. Empirically, they lead to very similar results as correspondence analysis in cases where both are applicable.

In the following I shall only discuss methods based on principal components or on multidimensional scaling. I shall illustrate them with graphs and shall give examples where some comparisons are possible, either between the recovered structure and the true underlying structure, or between methods, or both.

One should not expect wonders from any of these methods. They cannot give ready-made interpretations, and one should not expect anything like confidence levels from them. At best, they can give useful hints. The most convincing success stories have to do with seriation (recovery of an underlying linear order). But I have seen several successful, i.e. convincingly interpretable 2-dimensional, and even a few 3-dimensional layouts.

7.2   PRINCIPAL COMPONENT METHODS

These methods are based on on the truncated singular value decomposition. See Golub and Van Loan (1983) for the general background.

Let X be an $n \times p$ matrix, containing the coordinates of n points in p-space, ordinarily standardized in a suitable fashion, say centered at the mean. Then $X^T X$

is the covariance matrix. The singular value decomposition of X is defined as

$$X = USV^T,$$

where U is an orthogonal $n \times n$ matrix, S a diagonal $n \times p$ matrix, with $S_{ij} = 0$ for $i \neq j$, $S_{11} \geq S_{22} \geq \ldots \geq S_{kk} \geq \ldots \geq 0$, and V an orthogonal $p \times p$ matrix. This decomposition gives a representation of the covariance matrix: $X^T X = V S^T S V^T$, or $X^T X V = V S^T S$, from which it can be seen that the S_{jj}^2 are the eigenvalues of $X^T X$, and the column vectors of V are the eigenvectors.

We note that the matrices U and V define an orthogonal transformation of the original data, $Z = XV = US$, that maps the points in such a way that the leading coordinates Z_{i1} show the widest spread, since we have

$$\sum_i Z_{ij}^2 = S_{jj}^2.$$

We truncate this decomposition by putting $S_{kk} = 0$ for $k > m$. In data analytic applications one usually chooses $m = 2$ or $m = 3$. In the defining formula of the singular value decomposition the trailing columns of U and V thus are multiplied by 0, and by omitting the irrelevant columns of U and V we may write the resulting decomposition more parsimoniously in truncated form as

$$\hat{X} = \hat{U}\hat{S}\hat{V}^T,$$

where the matrices $\hat{U}$ and $\hat{V}$ contain the first m columns of U and V, thus $\hat{U}$, $\hat{S}$ and $\hat{V}$ are $n \times m$, $m \times m$ and $p \times m$, respectively. Note that $\hat{X}$ is the best possible approximation (in several respects) of X by a matrix of rank m.

The beauty of this approach is that it is self-dual: by transposing the defining formula:

$$X^T = V S^T U^T,$$

we get the singular value decomposition of the transposed matrix X^T. This way, we obtain a grouping of the columns of X, that is of the "variables" instead of the rows or "items". In terms of the truncated decomposition we have:

$$\hat{X}^T = \hat{V}\hat{S}\hat{U}^T.$$

Moreover, as $X\hat{V} = \hat{X}\hat{V} = \hat{U}\hat{S}$ and $X^T\hat{U} = \hat{X}^T\hat{U} = \hat{V}\hat{S}$, we may say that the matrices $\hat{V}$ and $\hat{U}$ map the "items" and the "variables", respectively, into the same m-space. Apart from the scaling by $\hat{S}$, the leading m columns of U and V give the coordinates of the images of the "items" and the "variables", respectively. The meaning of this will become clearer in the following example.

7.2.1   Principal component methods: Jury data

This example was chosen because it is small enough so that it can be used to illustrate both the analytical approach based on principal components and the intuitive approach, where one inspects the variables and interprets their conceptual meaning. If the number of variables is large, one will begin with the analytical approach. Then one inspects and interprets selected members of the analytical layout, and finally, one checks whether the interpretation extends also to neighboring members. The anecdotal beauty of this example is that two students, working independently of each other in an exam situation, one using the intuitive and the other the analytical approach, obtained essentially the same results.

The data for this example were taken from a preprint of a study by P. Ellsworth, B. Thompson, and C. Cowan on the "Effect of Capital Punishment Attitudes on Juror Perceptions of Witness Credibility". See Cowan et al. (1984) and Thompson et al. (1984).

In 1968 the U.S. Supreme Court ruled that jurors who make it unmistakably clear

(1) that they would automatically vote against the imposition of capital punishment without regard to any evidence that might be developed at the trial of the case before them, or

(2) that their attitude toward the death penalty would prevent them from making an impartial decision as to the defendant's guilt

cannot serve on capital case juries. Jurors excluded by the above standards are referred to as "excludable", all others as "qualified". It does not seem unreasonable to suspect that this rule gives rise to juries in which "hard liners" are over-represented and which are biased against the defendant.

In order to get some empirical evidence on this conjecture, P. Ellsworth, B. Thompson, and C. Cowan conducted the following experiment: They produced a movie of a fictitious court room scene, where a white police officer and a black defendant give their respective views of a confrontation that had occurred between them on a crowded sidewalk after a concert, late one evening in the summer. The officer, along with 11 other officers, was attempting to move the pedestrian traffic from the concert down a sidewalk. The defendant was moving up the sidewalk against the flow of traffic, in order to get to his car which was located just behind the officers. The defendant and an officer became engaged in a conversation which climaxed in a physical confrontation. The defendant was arrested for assaulting a police officer.

The movie was shown to a number of subjects, all of whom were eligible for jury duty in general, but some of whom were excludable and some were qualified. After seeing the movie, the subjects were asked to fill out a questionnaire (see Exhibit 7.1). Questions 1-3 measure subject's perception of the general accuracy and truthfulness of the two witnesses. Questions 4-10, 13, 14 measure which and to what extent specific facts mentioned in the witness' testimony are believed by the subject to be true. Questions 11, 12, 15, 16 measure the subject's sympathy for the witness' situation or position in this particular incident. In addition, demographic information on the subjects was recorded. The data set is listed and described in Exhibits 7.2 and 7.3; it gives the results of the experiment for 35 subjects.

1. In general, how truthful was Officer Jensen?

      1   2   3   4   5   6

   not at all   completely

2. In general, how truthful was the defendant, Marvin Johnson?

   1   2   3   4   5   6

   not at all   completely

3. Whose description do you think was more accurate?

   1   2   3   4   5   6

   defendant   Officer Jensen

4. Do you believe the defendant threatened to punch Officer Jensen?

   1   2   3   4   5   6

   no   yes

5. Do you believe the defendant struck Officer Jensen's chin?

   1   2   3   4   5   6

   no   yes

6. Do you think the Officer's chin was struck by a bottle?

   1   2   3   4   5   6

   no   yes

7. Do you think the defendant was limping when escorted to the police car?

   1   2   3   4   5   6

   no   yes

8. Do you think Officer Jensen ever used a racial slur when addressing the defendant?

   1   2   3   4   5   6

   no   yes

9. Do you think the defendant ever addressed the police officer in a derogatory fashion?

   1   2   3   4   5   6

   no   yes

10. Who do you think initiated the struggle that occurred?

    1   2   3   4   5   6

    defendant   officer

11. Was Officer Jenser more rough than necessary to control the situation?

    1   2   3   4   5   6

    no, not at all   yes, completely

12. Was the defendant justified in trying to break through the police line?

    1   2   3   4   5   6

    no, not at all   yes, completely

13. Do you think the officer was antagonistic toward the defendant before the struggle occurred?

    1   2   3   4   5   6

    no   yes

14. How likely is it that the defendant had been in trouble with the police before this incident?

    1   2   3   4   5   6

    not at all   very

15. To what extent did the officer treat the defendant unfairly because the defendant was black?

    1   2   3   4   5   6

    not at all   completely

16. To what extent was the defendant's behavior belligerent because of his attitude toward police officers?

    1   2   3   4   5   6

    not at all   completely

Exhibit 7.1   Jury data: Questionnaire. The subjects were asked to circle the number they felt most appropriate.

	1	2	3	4	5	6	7	8	9	10	11	12	13	14	15	16	17	18	19	20	21	22	23	24	25	26
1	1	5	2	5	6	6	1	3	4	6	1	2	2	4	5	1	5	23	0	3	7	0	1	2	1	
2	1	5	2	5	6	6	1	1	4	5	1	2	3	5	6	4	5	56	0	1	5	0	1	5	0	
3	1	4	3	4	4	3	1	3	5	5	4	4	3	4	3	4	4	55	0	2	8	0	1	4	1	
4	1	5	4	5	4	5	2	4	3	4	2	5	2	3	6	2	5	52	0	1	6	1	1	5	1	
5	1	3	3	2	1	1	6	6	6	4	3	5	2	2	3	2	4	64	0	3	8	0	1	1	1	
6	1	3	4	3	3	4	2	5	4	4	4	3	1	4	4	4	4	61	1	1	6	0	2	4	0	
7	1	5	2	5	6	5	1	4	3	5	2	4	2	4	5	3	5	58	1	1	8	0	1	1	0	
8	1	6	1	6	6	6	1	3	2	5	1	5	2	1	5	1	4	52	0	1	7	1	1	1	0	
9	1	5	2	6	6	6	1	4	4	6	1	1	1	1	6	1	6	36	0	5	6	0	1	1	1	
10	1	4	4	4	3	3	4	5	3	5	5	4	3	5	5	2	4	37	0	5	8	0	2	3	1	
11	1	5	3	5	4	6	1	5	6	6	1	1	1	5	6	4	6	57	0	4	4	1	1	2	1	
12	1	5	3	6	3	5	1	2	1	6	1	1	1	1	6	4	5	62	1	1	9	1	1	1	0	
13	1	5	2	6	4	6	1	3	4	3	1	1	1	1	4	2	2	34	1	1	3	0	2	1	0	
14	1	4	5	2	3	3	1	1	3	4	2	1	1	3	3	3	3	29	1	2	3	0	5	4	1	
15	1	4	4	2	2	2	4	3	6	6	4	4	3	3	4	2	4	30	0	2	5	0	2	4	1	
16	1	5	4	5	5	5	1	4	2	5	1	5	2	3	6	2	4	61	0	1	5	1	1	1	0	
17	1	6	3	4	6	6	1	1	1	6	1	1	1	3	6	2	6	67	1	1	6	0	2	1	0	
18	1	4	3	4	5	6	1	5	5	6	5	4	1	1	6	5	5	34	0	2	6	0	2	1	1	
19	1	3	4	3	2	2	4	3	3	4	5	6	6	5	3	2	4	28	1	1	5	0	1	4	1	
20	0	3	3	4	3	3	1	4	4	4	4	3	5	4	5	4	4	21	1	5	6	1	2	2	1	
21	0	3	5	1	1	1	1	1	6	6	6	6	4	6	1	4	2	48	0	1	7	0	2	5	0	
22	0	5	2	5	5	5	1	6	2	5	1	2	1	3	5	3	4	54	0	1	3	1	1	2	1	
23	0	4	3	4	3	2	5	5	4	5	2	3	2	5	2	4	4	31	0	1	5	0	5	5	1	
24	0	4	2	3	2	4	2	2	5	5	4	5	4	5	6	4	4	23	0	5	1	0	5	4	1	
25	0	2	5	2	2	2	4	5	5	2	5	6	6	5	2	4	2	23	1	5	7	0	5	4	1	
26	0	3	4	3	2	2	5	5	5	5	5	4	4	5	4	4	4	39	0	5	9	0	1	5	1	
27	0	3	3	3	3	2	5	3	5	5	2	4	2	4	5	3	5	68	0	1	6	1	2	2	1	
28	0	2	2	4	1	2	3	5	6	6	4	5	5	6	3	4	4	19	0	5	6	0	1	1	1	
29	0	4	4	5	5	4	1	5	4	5	3	2	5	4	4	3	4	28	1	5	9	1	5	4	1	
30	0	3	3	4	4	4	4	3	4	4	4	4	4	4	5	4	4	70	0	4	5	1	2	5	1	
31	0	5	4	5	5	1	1	1	4	5	5	4	1	1	2	1	5	64	0	2	4	1	1	1	1	
32	0	4	4	3	2	2	4	3	3	4	4	5	5	4	3	2	2	32	1	1	6	0	2	2	0	
33	0	3	5	2	5	1	5	4	6	5	6	6	6	6	5	1	4	49	0	2	3	0	2	2	1	
34	0	2	5	1	1	1	1	4	6	1	6	6	6	6	2	3	2	18	0	5	5	1	1	4	1	
35	0	5	5	3	2	2	2	3	2	2	4	2	4	2	2	2	2	34	0	2	6	1	2	5	1	

Exhibit 7.2   Jury data. See Exhibit 7.3 for the identification of the variables.

1) subject identification

2) group identity   0:   excludable

                                 1:    qualified

3)

···   answers to the questions

18)

19) age

20) sex   0:   female

                                            male

21) marital status   1:   married

                                 2:    divorced

                                 3:    separated

                                 4:    widowed

                                 5:    never married

22) education   1:   0-8 grades

                                 2:    some high school

                                 3:    high school complete

                                 4:    vocational/technical school

                                 5:    community college

                                 6:    4 yr college incomplete

                                 7:    4 yr college complete

                                 8:    some graduate work

                                 9:    masters degree

23) employment status   0:   employed

                                 1:    unemployed

24) political party   1:   Republican

                                 2:    Democrat

                                 3:    American Independent

                                 4:    Peace and Freedom

                                 5:    independent voter

                                 6:    other

25) religious preference   1:   Protestant

                                 2:    Catholic

                                 3:    Jewish

                                 4:    none or atheist

                                 5:    other

26) previous jury service   0:   yes

                                 1:    no

Exhibit 7.3   Jury data, identification of variables.

The problem here is not one of sorting out the subjects (the rows of the data matrix), but to group the questions (the columns 3 to 18) and to distill from them a common description of the hardline–softline attitude, independent of the classification as "qualified" or "excludable".

The singular value decomposition was applied to the data matrix (all 35 rows, and columns 3 to 18) centered at column means. The leading singular value is dominant; in the variance decomposition $\sum S_{jj}^2$ the three leading terms contribute 50%, 10% and 8%, respectively. Exhibit 7.4 (top) gives a map of the questions in the plane spanned by the first two principal components. One notices that the first principal component direction splits the questions into clusters. Note that the x-y-coordinates of the figure correspond to the first two columns of V multiplied by S_{11} and S_{22} respectively. The left hand group contains the questions 1, 3, 4, 5, 9, 14, 16, and the right hand group the questions 2, 6, 8, 10, 11, 12, 13, with two questions (5 and 7) somewhere in between. Upon examination of the questionnaire one realizes that for the left hand group the typical answer of a "hardliner" is on the high end of the multiple choices, while for the right hand group it is on the low end. If we center the answers to the questions by subtracting 3.5, multiply the answer to question i by V_{i1} and add them up, we obtain an "objective" combined softline–hardline attitude score.

Exhibit 7.4 (bottom) shows a map of the subjects. Note that the first principal component direction for each subject gives an algebraically weighted sum of the answers given by that subject, with coefficients taken from the first column of V, that is what we just have called the "objective" combined attitude score. Thanks to this weighting, most of the "qualified" subjects, supposedly the hardliners, migrate to the left hand side of the figure, the "excludable" to the right hand side. Thus in both graphs, we have an identifiable "hardliner" side on the left.

We used this example in the data analysis part of a Ph.D. qualifying exam at Harvard. These exams gave a palette of problems the students could choose from. They were asked to explain the purpose of the study, what questions they would want to answer and what questions they could answer with the data at hand. They should explore the data, decide on the appropriate techniques and check the validity of their results. But they were not given problem-specific hints. The two students selecting this particular problem realized that it would be advisable to combine the answers to the 16 questions into a single measure of "hardliner-tendency". One of them chose a common sense *a priori* approach, deciding that in one group of the questions (1, 3, 4, 5, 9, 14, 16) the high end of the answers would be typical for hardliners, while in another group (2, 6, 8, 10, 11, 12, 13) the low end would be typical, and that questions 7 and 15 were somewhat neutral. In other words, she obtained the same grouping as the "objective" one described above. So she aligned the answers by inverting the numbers for one of the groups and then combined the numbers by adding them, omitting questions 7 and 15. The other student decided on an analytical approach and on his own invented the one sketched above. He calculated principal component scores (the first column of our matrix V) and used them for weighting the answers, as we have done above. Of course, the "analytic" approach does not circumvent the intuitive interpretation – that the thus weighted "attitude score" indeed can be interpreted as a single softliner-hardliner score – but through the great discrepancy between the first two singular values it confirms that the questions appear to address one single attitude.

In short, through these approaches we have been able to combine the answers to the questions into a single variable describing the hard-soft attitude. But at this point the really hard part of the interpretative analysis only begins. One must address questions such as whether, and to what degree, the property of being qualified or excludable and the hard-soft attitude are conditioned on common demographic properties, such as religious preference, age or education.

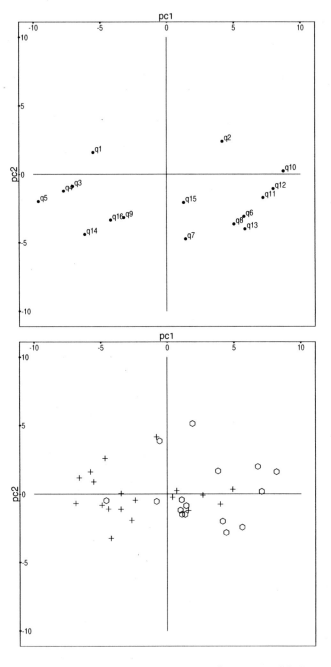

Exhibit 7.4 Jury data: principal component maps (questions: q, "qualified": +, "excludable": hexagon).

7.3 MULTIDIMENSIONAL SCALING

7.3.1 Multidimensional scaling: the method

Assume that you have n items and an $n \times n$ *similarity matrix* S_{ij}, expressing the "similarity" between items i and j. This similarity matrix may contain missing values. The similarities need not be expressed in numerical form, but a sufficient number of item pairs should be comparable in the sense that it is known whether items (i, j) are at least as similar to each other as items (k, l): $S_{ij} \succ S_{kl}$. The idea now is to find n points in a low-dimensional Euclidean space, such that the Euclidean distances d_{ij} between the points as far as possible match the similarity relations between the respective items:

$$S_{ij} \succ S_{kl} \implies d_{ij} \leq d_{kl}.$$

It is far from trivial to turn these ideas into a working computer program. The first successful implementation seems to be due to J. B. Kruskal (1964). Since then, a large number of such programs, with ever more options and variants have been created. Of course, the orientation of the resulting configuration is undefined; customarily, a principal component transformation is applied at the end. The examples below were produced using a slightly modified version of Kruskal's 1973 program KYST (the letters stand for Kruskal – Young – Shepard – Torgerson).

7.3.2 Multidimensional scaling: a synthetic example

Consider a 4×6 point-grid with integer-valued coordinates (x_i, y_i) and define the "Manhattan distance" between two points i and j as

$$d_{ij} = |x_i - x_j| + |y_i - y_j|.$$

Define a similarity matrix by

$$S_{ij} = \begin{cases} 1 & \text{for } d_{ij} \leq 2, \\ 0 & \text{otherwise.} \end{cases}$$

Actually, this similarity matrix does not give good results, but if we apply a standard trick and replace it by SS^T, it works. The result is shown in Exhibit 7.5.

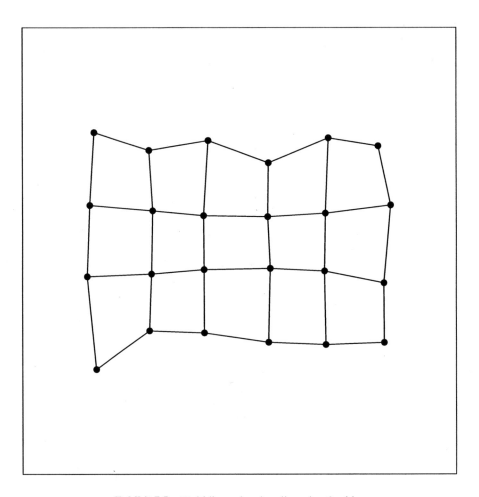

Exhibit 7.5 Multidimensional scaling: 4×6 grid.

7.3.3   Multidimensional scaling: map reconstruction

If the items are cities and the dissimilarity measure is, say, the distance in miles, or the travel time between two cities, then, as a rule, multidimensional scaling is able to reconstruct a fairly accurate geographical map. For an example see Kendall (1970).

However, this works only if the dissimilarity measure is somehow, possibly loosely, related to geographical distance. Attempts to reconstruct the geography of ancient Anatolia from written documents for example were unsuccessful. The number of times two cities occurred together in a document was taken as their measure of similarity. Unfortunately, this similarity measure does not correlate well with topography: in commercial documents the endpoints of trade routes seem to be more similar to each other than to intermediate stations.

On the other hand, applications to linguistics turned out to be highly useful. First of all, they facilitated the classification of linguistic data collected from a large number of dialects. Similarity between dialects was determined as follows. A native speaker was approached with a list of basic notions (usually consisting of 100 or 400 items, corresponding to everyday nouns, verbs, etc.) and asked how he or she would render them in his or her dialect. The measure of similarity between two dialects was the percentage of roots the lists had in common. A comparison of the map constructed from these lexicostatistical percentages through multidimensional scaling with the known geographical map turned out to be interesting. Multidimensional scaling disagreed with the geographical structure in a revealing fashion: it would group the coastal dialects close together. See Paul Black (1976).

7.4   CORRESPONDENCE ANALYSIS

7.4.1   Correspondence analysis: the method

Correspondence analysis was developed from the 1960s onward by the French statistician Jean Paul Benzécri (1992). It is an application of the singular value decomposition to large contingency tables.

I first describe the method by giving a recipe. Assume that X is a $I \times J$ matrix with non-negative elements (usually counts, often just taking 0-1 values). We begin by standardizing this matrix to a matrix Q. Standardize the total sum to 1:

$$P_{ij} = \frac{X_{ij}}{\sum_{ij} X_{ij}},$$

calculate row and column sums:

$$r_i = \sum_j P_{ij}, \quad c_j = \sum_i P_{ij},$$

and put

$$Q_{ij} = \frac{P_{ij} - r_i c_j}{\sqrt{r_i c_j}}.$$

The heuristic idea behind this standardization is: if the elements of X are independent Poisson variables with parameters $\lambda \rho_i \gamma_j$, where $\sum \rho_i = \sum \gamma_j = 1$, then r_i, c_j are estimates of ρ_i, γ_j, and the elements of Q are approximately independent with mean 0 and variance 1.

Now apply a singular value decomposition to Q:

$$Q = USV^T,$$

where U and V are orthogonal matrices, and S is a diagonal $I \times J$ matrix, with $S_{11} \geq S_{22} \geq \ldots \geq 0$. Truncate S to its top left $m \times m$ corner and the matrices U, V to their first m columns, where m is a small number (1 to 5). This gives new matrices $\hat{U}, \hat{S}, \hat{V}$. Note that the so-called truncated singular value decomposition $\hat{Q} = \hat{U}\hat{S}\hat{V}^T$ gives the best approximation (in several respects) of Q by a matrix $\hat{Q}$ of lower rank m. Now put:

$$F_{ik} = u_{ik} s_{kk} / \sqrt{r_i}, \quad G_{jk} = v_{jk} s_{kk} / \sqrt{c_j},$$

where k now ranges only from 1 to m. The matrices F and G are the mappings of the rows and columns of X into m-space, respectively. The purpose of the somewhat arbitrary final rescaling with $\sqrt{r_i}$ and $\sqrt{c_j}$ is to get a neater common mapping of f and g into the same m-dimensional space.

7.4.2 Kültepe eponyms

The following example was chosen because it offers a complete spectrum: the sequencing of a data set by hand, by correspondence analysis and by multidimensional scaling, and finally, after all this had been done, the true sequence was discovered.

Around 1900 BC, a large Old Assyrian trade colony, centered at the Karum Kanish (modern Kültepe, near Kayseri) flourished in Asia Minor. Several 10000 commercial records in the form of cuneiform tablets have been excavated. A fair number of the texts are dated by the names of certain high officials, who apparently held this office for one year each. Much later, the Romans used a similar system, namely dating the years after the two consuls in office. Until 1998 the chronological sequence of these officials, or eponyms, was not known.

In the 1970s M. T. Larsen had attempted a preliminary ordering of the year-eponyms based on 41 texts (mostly records of various loans) that mentioned more than one eponym, see Larsen (1976, p. 375-382). Eponyms mentioned together in the same text should also be chronologically close together. In the mid-1990s, I subjected Larsen's data to a correspondence analysis.

Then, in 1998, two texts listing the eponyms in sequence were discovered. Thanks to this so-called Kültepe Eponym List (KEL), published by Veenhof (2003), it is now possible to compare both Larsen's ordering and the ordering produced by correspondence analysis to the true ordering as given by KEL.

The tables below give the incidence matrix: each of the 38 rows corresponds to an eponym, each of the 41 columns to a cuneiform tablet, and the matrix records a 1 if the eponym occurs on the tablet, a dot otherwise. Larsen ordered the rows and columns in such a way that the entries 1 were made neighbors as far as feasible.

Instead of the actual names of the eponyms, given by Larsen, the tables below give their KEL-numbers, to show their true sequence (still unknown to Larsen). Thus, for example, KEL076 corresponds to the eponym *Sallija*. We know now that the very first entry in Larsen's list, here denoted KEL000, was not an eponym. Already Larsen had expressed doubts, and the name does not occur in KEL. It emerges from the list that Larsen's 37 real eponyms coincide with the 37 names KEL076, KEL079 to KEL114, although in a somewhat different order. It seems that this time segment of 39 years, from KEL076 to KEL114, roughly coincides with the time when the trade colony flourished.

Here is the 38×41 incidence matrix (38 eponyms, 41 texts), in Larsen's ordering:

```
------------------------------------------------------------
KEL000  1.......................................
KEL076  .111....................................
KEL087  111.11..................................
KEL084  ....1111................................
KEL080  ....111.1...............................
KEL083  1..1111.1111............................
KEL090  ...1.1..1.1.............................
KEL085  ....111.1.1.1...........................
KEL091  ...111..1..1.11.........................
KEL092  ...1.......1111.........................
KEL081  .......1.......1........................
KEL086  .......1.1......1.......................
KEL089  .....1....11.....1......................
KEL097  .....1...1.....1..1.....................
KEL088  .......1........1.......................
KEL082  .......1..........1.....................
KEL094  ............1.1..1......................
KEL096  ............11...1......................
KEL095  ............1...........................
KEL093  ............1.....1.....................
KEL079  ................1.......................
KEL114  ..................11....................
KEL098  ........1........11..11.................
KEL101  ........1.......1......111..............
KEL102  ................1...11111.1.............
KEL103  ......................11.11.............
KEL104  ..................1...11.111............
KEL109  ...................1.........11.........
KEL105  ...........................111..........
KEL100  .........................11....1........
KEL106  ............................1.11111...
KEL108  .....................1......11111..1..
KEL107  .....................1......1.11111.1.
KEL112  ................................1.......1..
KEL099  ...........................1............
KEL113  ......................................11
KEL111  .......................................1
KEL110  .......................................1
```

The incidence matrix in correspondence analysis order (both the rows and the columns have been reordered):

```
-------------------------------------------------------------
KEL076 11...1...................................
KEL000 ..1.....................................
KEL087 1111..1.................................
KEL084 ...11.11................................
KEL080 ...11.1....1............................
KEL090 .....11.1..1............................
KEL085 ...11.1.1..11...........................
KEL081 .......1.1..............................
KEL083 ..11111.1.11......1.....................
KEL091 ...1.11...11..11........................
KEL092 .....1......1111........................
KEL086 .......1.1........1.....................
KEL089 .....1.1.1........1.....................
KEL082 ...........1.....1......................
KEL097 .....1......1..1.1......................
KEL096 ............111.........................
KEL094 ...........1.1....1.....................
KEL095 ...............1........................
KEL088 ..........1......1......................
KEL079 ...............1........................
KEL098 ............111.11......................
KEL093 ...............1.........1..............
KEL101 ..................11...11...1...........
KEL102 ..................1111.1.1.1............
KEL114 ........................1..1............
KEL099 ............................1...........
KEL103 ........................1.111...........
KEL100 ........................1..1..1.........
KEL104 ........................11111.1.........
KEL109 ..........................1.....1.1.....
KEL112 ............................1.1.........
KEL105 ............................11.1........
KEL108 ...........................11..11111....
KEL107 ...........................1....1111111.
KEL106 ..............................111111..
KEL113 .......................................11
KEL111 ........................................1
KEL110 ........................................1
```

Compare rankings: true, Larsen, Correspondence analysis, and their differences.

| KEL- | ranks: | | | differences: | | |
number	true	Larsen	CorAn	Larsen-true	CA-true	CA-Larsen
KEL000	1	1	2	0	1	1
KEL076	2	2	1	0	-1	-1
KEL079	3	21	20	18	17	-1
KEL080	4	5	5	1	1	0
KEL081	5	11	8	6	3	-3
KEL082	6	16	14	10	8	-2
KEL083	7	6	9	-1	2	3
KEL084	8	4	4	-4	-4	0
KEL085	9	8	7	-1	-2	-1
KEL086	10	12	12	2	2	0
KEL087	11	3	3	-8	-8	0
KEL088	12	15	19	3	7	4
KEL089	13	13	13	0	0	0
KEL090	14	7	6	-7	-8	-1
KEL091	15	9	10	-6	-5	1
KEL092	16	10	11	-6	-5	1
KEL093	17	20	22	3	5	2
KEL094	18	17	17	-1	-1	0
KEL095	19	19	18	0	-1	-1
KEL096	20	18	16	-2	-4	-2
KEL097	21	14	15	-7	-6	1
KEL098	22	23	21	1	-1	-2
KEL099	23	35	26	12	3	-9
KEL100	24	30	28	6	4	-2
KEL101	25	24	23	-1	-2	-1
KEL102	26	25	24	-1	-2	-1
KEL103	27	26	27	-1	0	1
KEL104	28	27	29	-1	1	2
KEL105	29	29	32	0	3	3
KEL106	30	31	35	1	5	4
KEL107	31	33	34	2	3	1
KEL108	32	32	33	0	1	1
KEL109	33	28	30	-5	-3	2
KEL110	34	38	38	4	4	0
KEL111	35	37	37	2	2	0
KEL112	36	34	31	-2	-5	-3
KEL113	37	36	36	-1	-1	0
KEL114	38	22	25	-16	-13	3

The rank correlations are:

true ranks to Larsen:	0.865
true ranks to correspondence analysis:	0.890
Larsen to correspondence analysis:	0.978

We note that correspondence analysis gives an agreement with the true order that is just a tad better than Larsen's ordering by eye and hand. Both Larsen and correspondence analysis erred sorely with the positions of KEL079 and KEL114. But the agreement between correspondence analysis and Larsen is very close. Interestingly, correspondence analysis reached a good agreement with the true order even though the leading singular values of the standardized incidence matrix are very close together: the leading four S_{jj}^2 contribute 9.3%, 8.7%, 7.5% and 7.4%, respectively, to $\sum S_{jj}^2$.

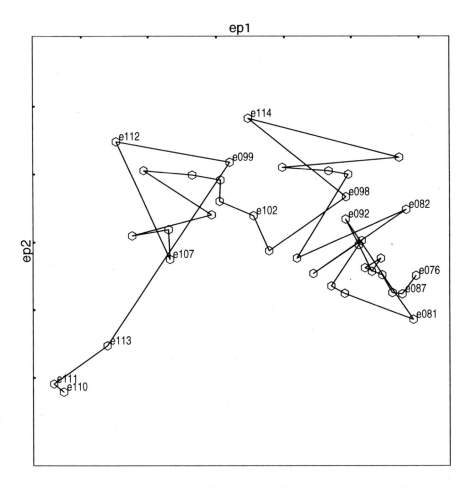

Exhibit 7.6 Multidimensional scaling: Eponym data. The eponyms are connected in Larsen's order.

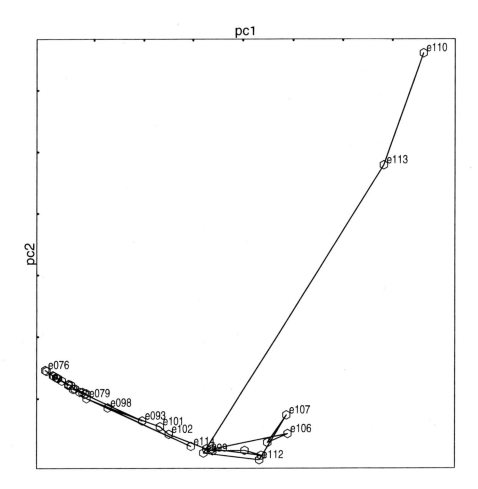

Exhibit 7.7   Correspondence analysis: Eponym data. The eponyms are connected in Larsen's order.

7.4.3 Further examples: marketing and Shakespearean plays

In my opinion, some of the most fascinating applications of correspondence analysis occur in marketing research. I cannot present a case study, because on one hand the data sets in question are proprietary, and on the other hand, they are too large and their investigation is too involved. But I can attempt to give a rough description.

A typical problem in marketing is that you have a new product, and you would like to find out in which magazines to advertise and what types of ads to use in order to reach, and to appeal to, the potential buyers of that product. For that, it helps to map the population of potential customers according to demographic characteristics (age, sex, education, income, ...), attitudes (conservative, progressive, pessimistic, optimistic, ...), preferences (journals, types of pictures, ...). Correspondence analysis was used successfully to produce interpretable two- or three-dimensional spreads of such data. Then one has to identify the region of the data landscape where the potential buyers sit, what journals are prevalent in that corner, and what kinds of ads are likely to appeal to those people.

In a case I am familiar with, the intuitive interpretation of the two leading dimensions of a correspondence analysis map of attitudes was that one coordinate ranged from extroverted to introverted, and the other coordinate from conservative to progressive. A fascinating, as yet unpublished result of such studies was recently discussed in the weekly *Die Weltwoche* (2 December 2010). The study in question, conducted by the marketing research institute Demoscope, shows a surprising trend reversal of the attitudes of the Swiss population at the beginning of the decade. From 1974 (when the surveys began) to the end of the century, the popular attitudes moved steadily from the conservative toward the progressive pole. Then a unique change of direction occurred, and between 2001 and 2010, the move was back toward conservative.

Incidentally, in some of these marketing studies one also applied methods described in Section 8.3 (comparison of point configurations) to force the results of studies that had been done with the same questionnaire, but with different subject populations, into comparable orientations.

Some of the relevant issues can be illustrated with the help of a simple example from the humanities, taken from Brainerd (1979). The particular problem was to find linguistic "keys" to genre. Brainerd used discriminant analysis and found a significant relationship between the use of personal pronouns and genre. In the following I shall present a more intuitive, but highly revealing analysis through correspondence analysis.

For each of the 38 plays of Shakespeare, the following information is given (see Exhibit 7.8):

- the traditional classification into genres (Comedy, History, Tragedy, Romance). Troilus and Cressida is ambiguously classified first T, later C.

- the total number of words.

- the number of personal pronouns (1s, 1p: first person singular/plural; 2s, 2p: second person singular/plural; 3m, 3f, 3n: third person singular, masculine, feminine, neuter; 3p third person plural; totp: total number of personal pronouns).

- the order of plays in the first folio edition (presumably the number in which the plays were written).

Correspondence analysis (using the eight columns with pronoun counts, from 1s to 3p) gave two large leading singular values: the two leading terms of $\sum S_{jj}^2$ contribute 41% and 34%, respectively, and the resulting 2-dimensional map (see Exhibit 7.9) is interesting and suggests some intriguing interpretations. In the space spanned by the cases (i.e. plays), the Comedy genre, with Romance behind, migrated to the north-east corner, while History, with Tragedy behind, migrated to south-west. There were only few exceptions (e.g. Othello) and little overlap. In the space spanned by the variables (i.e. pronouns), the north-east corner corresponds to third person feminine pronouns (3f). That corner collects mainly the Comedy and Romance genre, presumably with women in leading roles, while the south-west corner collects mainly the Tragedy and History genre, presumably mostly with male protagonists. The other diagonal does not seem to be directly relevant for the traditional notion of genre; it seems to distinguish between 1s and 2s towards north-west, that is first and second person singular pronouns (much dialogue?), and plural and neutral pronouns towards south-east (little dialogue?). Thus, while correspondence analysis does not give final results, it can give hints where to look for them.

No.	Play	Genre	totw	1s	1p	2s	2p	3m	3f	3n	3p	totp	order
1	Tempest	R	16036	998	167	401	329	283	34	200	135	2548	36
2	TwoGentlemen	C	16883	1164	93	340	513	304	182	244	87	2927	3
3	MerryWives	C	21119	1445	146	165	773	552	92	280	121	3575	18
4	Measure	C	21269	1029	176	226	842	609	60	392	89	3424	26
5	ComedyErrors	C	14369	1062	129	274	432	272	57	180	66	2473	2
6	MuchAdo	C	20768	1262	135	220	700	488	155	298	136	3394	19
7	LovesLabor	C	21033	930	216	236	534	345	53	280	122	2716	1
8	Midsummer	C	16087	873	165	251	409	235	64	143	65	2206	11
9	Merchant	C	20921	1364	133	234	647	450	43	272	90	3234	13
10	AsYouLikeIt	C	21305	1240	149	313	740	550	63	262	106	3424	21
11	TamingShrew	C	20411	1287	163	300	718	354	149	287	69	3328	14
12	AllsWell	C	22550	1376	253	292	734	694	117	423	125	4014	23
13	TwelfthNight	C	19401	1271	107	343	645	500	85	300	63	3315	22
14	WintersTale	R	24543	1384	259	290	768	511	119	456	154	3942	35
15	KingJohn	H	20386	913	272	397	420	377	21	200	79	2679	12
16	RichardII	H	21809	1201	298	458	295	499	4	211	97	3064	10
17	HenryIV1	H	23955	1305	296	480	492	532	15	256	146	3522	15
18	HenryIV2	H	25706	1351	268	353	776	613	26	311	130	3829	16
19	HenryV	H	25577	944	460	251	623	635	26	313	190	3442	17
20	HenryVI1	H	20515	989	285	457	368	390	44	135	109	2778	4
21	HenryVI2	H	24450	1264	244	543	399	590	26	251	159	3476	5
22	HenryVI3	H	23295	1206	359	575	355	562	34	209	112	3412	6
23	RichardIII	H	28309	1745	300	568	689	609	55	269	154	4389	7
24	HenryVIII	H	23325	1246	214	77	803	673	56	320	202	3591	37
25	TroilusCress	?	25516	1113	288	346	586	693	93	307	133	3560	24
26	Coriolanus	T	26579	1041	503	276	986	935	15	391	290	4437	31
27	TitusAndron	T	19790	1003	231	519	380	357	78	188	151	2908	9
28	RomeoJuliet	T	23913	1305	176	605	414	376	134	289	95	3395	8
29	TimonAthens	T	17748	904	191	513	439	498	10	301	146	3002	32
30	JuliusCaesar	T	19110	947	275	230	562	528	7	231	133	2914	20
31	Macbeth	T	16436	693	285	222	339	370	22	248	132	2312	29
32	Hamlet	T	29551	1399	370	272	821	737	55	621	177	4452	25
33	KingLear	T	25221	1417	269	550	709	595	68	315	126	4050	28
34	Othello	T	25887	1638	157	320	728	594	181	522	121	4261	27
35	AnthonyCleo	T	23742	1209	404	423	539	643	91	377	131	3817	30
36	Cymbeline	R	26778	1451	350	406	687	622	131	435	140	4223	34
37	Pericles	R	17723	841	254	231	568	343	111	232	113	2694	33
38	TwoNobleKins	R	23403	1151	388	223	640	516	121	309	202	3551	38

Exhibit 7.8 Shakespeare's use of personal pronouns. From B. Brainerd (1979).

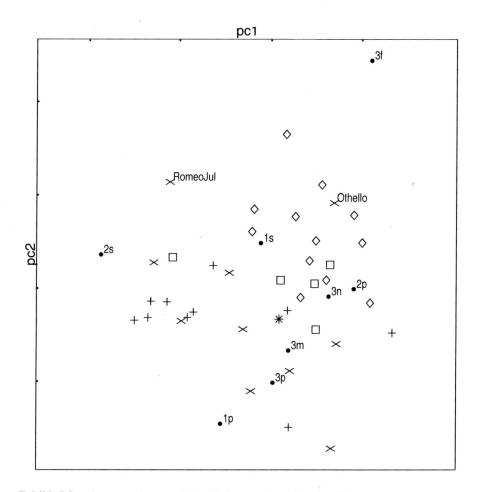

Exhibit 7.9 Correspondence analysis: Shakespeare's works. (+: History, ×: Tragedy, □: Romance, ◇: Comedy; *: Troilus and Cressida)

7.5 MULTIDIMENSIONAL SCALING VS. CORRESPONDENCE ANALYSIS

Where both multidimensional scaling and correspondence analysis are applicable, and both are used to map n items into an m-dimensional image space, they usually give similar results. However, it seems that the latter tends to give "cleaner" pictures. The likely reason is that the latter puts the noise predominantly into the invisible orthogonal complement of the image space, while the former forces much of it into the image space itself. Moreover, multidimensional scaling maps a truly linear structure into an open circle (called 'horseshoe' by Kendall), which then needs to be straightened out in order to obtain a sequencing, and with noisy data this can be tricky. The results of a correspondence analysis is more straightforward to deal with since it maps truly linear structure into a parabola rather than into an open circle, and the ordering can be read off from the first coordinate, see Exhibit 7.10. The algorithms for correspondence analysis are deterministic, and repeated analyses of the same data always give the same result, while multidimensional scaling starts with a random configuration and the optimization process may end up in different solutions. Whether this is an advantage or disadvantage is a matter of opinion. Also, multidimensional scaling has a broad palette of choices for the similarity measure and for dealing with it, which might be considered confusing by some and an advantage by others.

Exhibit 7.10 shows the results of multidimensional scaling and correspondence analysis, applied to an idealized version of the eponym data of Section 7.4.2, that is to a 38×41 matrix of zeros and ones clustering around the diagonal, with approximately six ones in each row and column, but without zeros in between.

The following subsections compare the two procedures with data sets that originally had provided prime examples for multidimensional scaling.

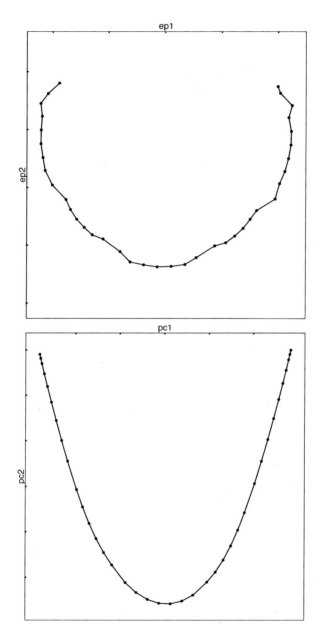

Exhibit 7.10   Synthetic linear data: Horseshoe (multidimensional scaling) and parabola (correspondence analysis).

7.5.1 Hodson's grave data

This data set has been treated in detail by D. G. Kendall (1970, 1971). The set derives from 59 iron age graves described by Hodson (1968). There were 70 different varieties of finds, fibulae, anklets, bracelets, etc., numbered from 1 to 70. See Exhibit 7.11 for a listing of the data. Kendall's results are shown in Exhibit 7.12.

I subjected the data to my modified version of Kruskal's 1973 KYST. First, the data were transformed into a 59×70 incidence matrix X. Then, multidimensional scaling was applied to the similarity matrix $X X^T$. The results are shown in Exhibit 7.13.

```
 1 │  1   2
 2 │  1   2   3
 3 │  1   2
 4 │  2   3   5
 5 │  4   5
 6 │  2   3   4   6
 7 │  2   3   4   7
 8 │  2   3   4   5   6   8
 9 │  2   4   6
10 │  3   9
11 │  2   3   5   8   9  10  11  12  13
12 │  3   4  12  13
13 │  2   3   6  10  14  15  16
14 │  2   3  17  19
15 │ 14  17  18  19  20  21
16 │ 14  17  18  22  23
17 │ 14  17  19  20  24
18 │ 23  25
19 │ 15  16  17  18  22  24  25
20 │ 21  24
21 │ 19  22
22 │ 22  26  27
23 │ 27  28  29  30  31  32
24 │ 27  28  29  30  31  32  33  34
25 │ 28  33
26 │ 11  23  27  29  30  31  33  34  35
27 │ 30  35
28 │ 22  36
29 │ 27  31  32  33  34  35  36  37
30 │ 27  33  34  35  36  38  39
31 │ 17  23  36  39
32 │ 36  37  38
33 │ 35  37  38  40
34 │ 37  38
35 │ 34  36  39
36 │ 27  34  36  40  41  42
37 │ 27  34  36  40  41
38 │ 34  40  42  43  44  45  46
39 │ 34  36  42  43  44  45  46  47
40 │ 34  36  37  47  48
41 │ 36  42  47  49  50
42 │ 34  36  37  44  45  47  48  49  51  52  53  54
43 │ 36  45  51  52  53  55
44 │ 36  50  51  54  55  56  57  58
45 │ 36  37  44  45  46  51  54  55  56  57  58  59  60
46 │ 54  57  61  62
47 │ 36  45  54  61  62  63
48 │  1  62
49 │ 62  64
50 │ 37  43  54  61  62  64  65
51 │ 61  63  66  67
52 │ 63  65
53 │ 60  63  66  68
54 │ 36  65  67  68
55 │ 65  66  67  68
56 │ 63  65  66  68  69
57 │ 66  69  70
58 │ 63  70
59 │ 66  70
```

Exhibit 7.11   Hodson's grave data: 59 graves, 70 different varieties of finds. From D. G. Kendall (1971).

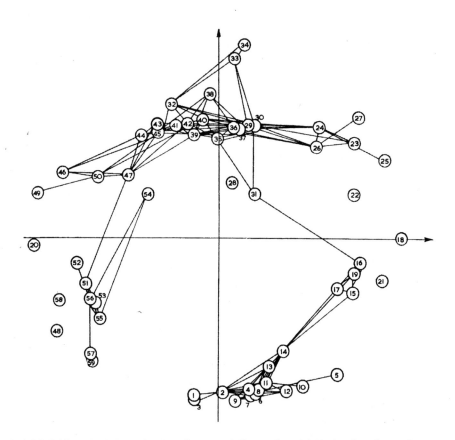

Exhibit 7.12   Multidimensional scaling: Kendall's results, with Hodson's order used to label tombs. Connecting lines indicate high similarity. Reproduced with permission, from D. G. Kendall (1970: p. 131, Fig. 3).

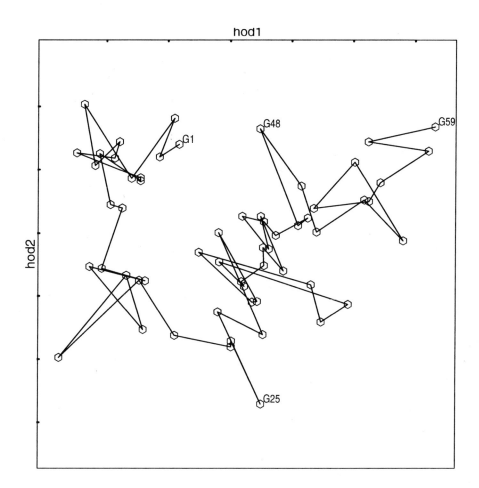

Exhibit 7.13   Multidimensional scaling: Hodson data. Connecting lines correspond to Hodson's order.

Then, the incidence matrix was subjected also to correspondence analysis. The results are shown in Exhibit 7.14.

We note the general similarity of the results: the graves are ordered roughly in the order of the archaeological seriation carried out by Hodson. Grave 48 is somewhat of an outlier, particularly in the correspondence analysis. Furthermore, the order also roughly corresponds to the geographical order of the graves in the elongated cemetery, see Exhibit 7.15 taken from Kendall (1970).

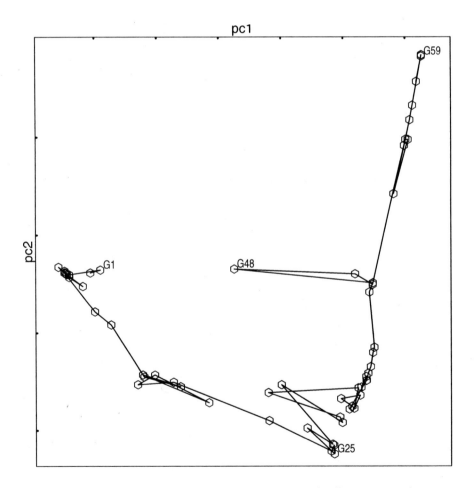

Exhibit 7.14 Correspondence analysis; Hodson data. Connecting lines correspond to Hodson's order.

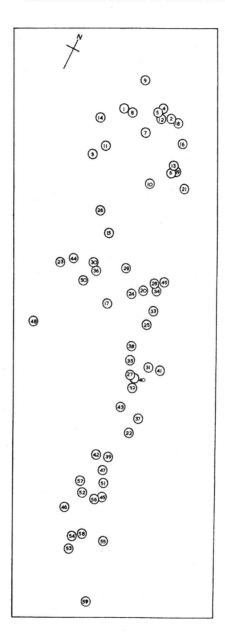

Exhibit 7.15 Hodson data. A true map of the Münsingen-Rain cemetery, in which the tombs (shown as circles) carry labels indicating their serial positions in the computer chronology derived from the 'horse-shoe' in Exhibit 7.12. Reproduced with permission, from D. G. Kendall (1970: p. 132, Fig. 4).

7.5.2   Plato data

This is a rather famous data set. The problem is to put the works of Plato into their proper chronological order. There is general agreement that the *Republic* comes at the early end, and so do most of the dialogues, while the *Laws* must be positioned at the late end. However, the position of five dialogues is uncertain (*Critias, Philebus, Politicus, Sophistes, Timaeus*).

Plato's rhythmical preferences show considerable evolution. According to Cicero the sentence endings (the so-called clausulae) are rhythmically the most important, and it had been conjectured a long time ago that Plato's rhythmical preferences were expressed in preferring some and neglecting other clausulae. This suggested to approach the problem by investigating the clausulae. The clausula was taken to consist of the last five syllables. Since there are two possible types of syllables (long and short), there are $2^5 = 32$ distinct classes of clausulae. The philologist Kaluscha (1904) was the first to use statistical methods in an attempt to assign an order to these works. His attempt found a positive echo only half a century later, in Brandwood's thesis and in the paper by Cox and Brandwood (1959). This work was based on Kaluscha's idea, but improved upon the statistical approach.

Kaluscha's data have been subjected to multidimensional scaling by L. I. Boneva (1971). Her analysis split the two longest works (the *Republic* and the *Laws*) into 10 and 12 separate books, respectively. The data used by Boneva are shown in Exhibits 7.16 and 7.17. The types of the clausulae are given on top, with ⌣ standing for short and − for long syllables.

The analysis presented here uses Boneva's data. I first subjected it to multidimensional scaling, using a modified version of Kruskal's 1973 KYST. The clausulae counts were standardized to row sum 1, and for the dissimilarity measure between two rows (that is, between two works of Plato) I took Euclidean distance.

The result of multidimensional scaling is shown in Exhibit 7.18. It shows a clear, albeit somewhat noisy, horseshoe structure.

Afterwards, the same data were subjected to a correspondence analysis. The results are shown in Exhibit 7.19.

Kaluscha had concluded that the order of the last six works of Plato is: *Timaeus, Critias, Sophistes, Politicus, Philebus,* and *Laws*. The computer analyses reach a similar result but interchange the positions of *Critias* and *Sophistes*.

The neat feature of correspondence analysis is that it maps the rows and the columns into the same space, which facilitates the interpretation of either. The map of the columns (i.e. of the clausulae) is shown in Exhibit 7.20, drawn in the same scale as Exhibit 7.19. Unfortunately, there is some overlap in the printed picture, but on closer inspection it appears that the left hand side (corresponding to the early works) prefers clausulae with a high number of changes between short and long syllables, while the right hand side (the late works) seems to prefer low numbers. In order to check this, I calculated the average number of changes for each of the works. The results are shown in Exhibit 7.21, where the works are ordered in reverse order of the average number of changes. This order agrees quite well with the order obtained from multidimensional scaling (Exhibit 7.18) or from correspondence analysis (Exhibit 7.19). The correlation between the two columns of Exhibit 7.21, that is, between the scores obtained from the average number of changes and from correspondence analysis, is -0.92, and the rank correlation is -0.88. But it seems that neither the average number of changes nor the first coordinate of the correspondence analysis gives the full picture: the second coordinate of the 2-dimensional representations apparently contains additional discriminatory information.

	⏑	—	⏑	⏑	⏑	⏑	—	—	—	—	⏑	⏑	⏑	⏑	⏑	⏑
	⏑	⏑	—	⏑	⏑	⏑	—	⏑	⏑	⏑	—	—	—	⏑	⏑	⏑
	⏑	⏑	⏑	—	⏑	⏑	⏑	—	⏑	⏑	—	⏑	⏑	—	—	⏑
	⏑	⏑	⏑	⏑	—	⏑	⏑	⏑	—	⏑	—	⏑	⏑	—	—	—
	⏑	⏑	⏑	⏑	⏑	—	⏑	⏑	⏑	—	⏑	⏑	—	⏑	—	—
charmides	5	4	6	8	11	6	4	9	10	10	14	17	17	8	20	11
laches	4	7	4	1	8	5	8	8	12	16	14	6	16	6	8	11
lysis	3	7	10	11	17	14	11	13	5	6	15	17	15	4	15	17
euthyphron	10	7	3	5	14	3	2	8	5	10	7	6	10	4	10	9
gorgias	16	18	24	19	22	27	22	21	30	42	36	29	38	27	39	44
hippias	2	5	2	8	4	6	1	3	4	5	7	10	6	1	9	15
euthydemus	8	11	10	11	13	9	9	17	20	12	12	35	21	21	37	24
cratylus	29	27	17	17	23	26	25	21	18	22	38	28	29	13	25	42
meno	5	4	8	9	11	5	5	6	13	25	21	18	23	8	14	11
menexenus	1	1	6	1	3	3	3	6	4	6	5	5	17	6	9	0
phaedrus	9	13	17	18	10	9	14	29	25	22	18	23	33	10	28	23
symposium	10	11	11	16	27	11	12	10	25	20	19	16	28	13	33	26
phaedon	17	16	15	10	21	15	16	19	10	41	15	21	38	23	32	30
theaetetus	19	23	18	20	21	20	16	23	38	28	39	32	25	29	39	35
parmenides	22	18	17	9	19	15	14	13	25	12	15	21	26	7	23	31
protagoras	9	11	12	10	14	13	8	11	21	9	20	14	15	15	11	8
crito	2	0	1	1	4	2	2	3	5	3	5	6	4	4	5	6
apology	2	3	8	2	11	10	6	5	14	10	7	11	14	11	13	13
R1	12	15	10	10	10	11	11	7	13	17	28	15	27	13	15	9
R2	3	9	8	7	9	6	16	7	6	12	12	9	23	11	13	11
R3	2	2	6	6	4	2	7	9	9	13	15	7	10	14	19	12
R4	3	3	4	4	9	13	10	8	10	17	10	7	24	7	16	10
R5	5	10	13	13	9	13	9	9	17	19	19	9	22	10	19	14
R6	5	7	5	10	9	8	4	10	13	19	7	12	16	19	13	11
R7	3	3	6	6	6	8	9	10	5	24	8	12	15	3	16	8
R8	2	1	5	5	9	5	3	6	12	17	5	9	11	5	23	9
R9	1	5	2	1	6	4	3	8	12	17	13	6	16	7	19	3
R10	5	5	6	10	8	4	7	7	8	19	9	12	9	11	14	6
L1	6	5	11	5	13	13	16	5	3	25	14	4	3	6	10	21
L2	6	7	4	7	10	11	5	6	0	17	10	3	4	3	3	20
L3	8	16	6	8	10	19	6	4	2	37	13	3	7	7	11	11
L4	7	13	4	10	12	11	9	9	3	20	7	3	2	5	9	17
L5	8	11	6	11	9	14	9	5	1	19	5	2	3	1	8	15
L6	9	22	4	6	12	20	11	4	1	30	15	1	1	9	13	32
L7	13	13	9	10	13	17	11	12	3	43	12	7	1	7	11	24
L8	4	8	3	8	10	10	3	5	1	23	3	2	1	3	8	10
L9	10	17	4	7	6	7	7	7	3	31	15	3	5	3	11	19
L10	4	13	8	14	9	9	8	4	2	40	11	3	6	4	9	21
L11	8	7	4	6	3	6	9	2	1	20	13	1	3	5	8	13
L12	7	12	8	8	8	8	8	4	2	28	12	4	5	5	12	12
critias	5	3	3	2	10	6	5	3	2	9	4	4	3	4	5	10
philebus	24	27	20	25	38	46	41	14	7	62	64	6	7	30	18	52
politicus	13	19	24	20	25	22	25	18	3	31	41	7	8	24	23	33
sophistes	26	33	31	24	22	23	30	37	19	21	30	15	28	28	28	47
timaeus	18	30	46	14	26	27	26	26	13	25	26	17	20	23	17	30

Exhibit 7.16   Plato data, clausulae 1-16.

	—	—	—	—	—	—	—	⌣	⌣	⌣	⌣	⌣	—	—	—	—
	—	—	—	⌣	⌣	⌣	⌣	—	—	—	—	⌣	—	—	—	—
	—	⌣	⌣	—	—	⌣	—	⌣	⌣	⌣	—	—	—	—	⌣	—
	⌣	—	⌣	⌣	—	—	—	—	⌣	—	—	—	—	—	⌣	—
	⌣	⌣	—	⌣	—	—	—	—	⌣	—	—	—	—	—	⌣	—
charmides	9	8	17	16	31	14	15	9	19	6	20	20	21	23	8	19
laches	10	22	9	7	22	13	13	9	19	6	24	18	12	30	8	20
lysis	13	16	11	9	19	16	6	10	19	3	10	5	8	14	4	4
euthyphron	5	5	9	5	9	12	11	10	9	3	9	12	10	11	3	9
gorgias	39	32	59	34	61	52	45	40	73	24	42	58	43	46	37	46
hippias	2	8	2	5	11	11	6	7	8	3	11	7	5	9	6	9
euthydemus	16	20	32	12	51	20	27	25	26	18	28	21	26	36	16	28
cratylus	27	52	22	17	35	36	32	34	47	9	39	25	31	34	21	38
meno	14	13	19	16	26	24	27	23	22	12	24	17	15	18	11	20
menexenus	6	4	8	4	4	6	6	6	8	3	11	6	6	14	9	7
phaedrus	20	17	39	18	29	42	21	40	26	9	26	19	37	39	11	19
symposium	18	26	23	26	48	29	22	39	35	22	27	24	25	45	13	32
phaedon	26	28	28	18	44	43	30	34	32	15	30	34	33	47	11	32
theaetetus	42	34	39	37	36	43	40	35	37	17	37	45	37	55	18	49
parmenides	13	19	24	13	24	28	15	30	16	9	26	18	26	21	7	25
protagoras	12	18	25	17	31	17	21	19	16	13	24	25	22	28	13	20
crito	3	3	6	6	9	6	6	9	10	3	8	5	7	13	1	10
apology	6	16	16	16	22	17	12	13	13	17	11	15	11	14	9	15
R1	11	11	11	7	30	26	22	20	23	10	25	14	18	18	13	29
R2	15	13	12	5	20	10	12	18	13	5	19	12	15	13	10	15
R3	21	10	11	19	25	12	13	13	19	19	15	13	18	22	4	22
R4	14	6	11	8	22	18	11	12	18	11	11	11	18	6	6	17
R5	14	15	19	7	27	23	15	18	18	10	15	16	18	22	11	18
R6	8	13	16	6	22	16	7	17	21	5	6	18	17	10	6	17
R7	6	6	18	7	29	14	8	15	15	3	5	16	12	21	11	6
R8	8	11	7	5	21	15	6	18	15	7	8	25	12	9	7	10
R9	6	16	11	5	21	16	8	13	15	11	12	8	9	20	4	9
R10	6	13	11	8	23	9	5	15	23	8	15	19	19	15	5	15
L1	13	8	3	13	8	3	15	4	24	5	15	13	12	27	8	10
L2	11	6	2	8	6	3	5	1	17	1	19	6	10	26	8	21
L3	15	5	1	14	5	3	8	3	22	7	14	8	5	22	7	12
L4	15	5	4	1	6	2	3	5	22	5	8	15	2	17	17	13
L5	13	3	1	3	9	1	5	0	18	5	3	10	6	15	13	11
L6	12	3	8	7	5	2	14	3	29	5	15	16	5	22	9	22
L7	21	5	1	6	13	3	20	6	46	9	17	13	11	32	12	29
L8	11	6	3	3	4	1	11	3	21	6	17	14	6	28	4	12
L9	8	2	2	4	9	1	7	6	36	9	15	14	8	28	16	13
L10	19	8	3	9	11	3	8	2	28	3	9	9	7	46	8	20
L11	7	2	7	9	7	0	5	5	17	13	13	9	4	40	5	15
L12	12	0	3	9	7	1	9	6	29	5	11	12	4	29	7	17
critias	4	3	1	3	2	7	2	4	8	5	3	7	9	3	5	6
philebus	53	7	4	11	27	7	25	12	51	32	32	32	23	86	28	47
politicus	53	21	5	26	14	6	35	8	34	19	29	38	16	52	22	56
sophistes	48	24	21	34	19	28	31	12	42	23	27	32	38	43	24	31
timaeus	23	25	25	25	23	21	23	25	23	17	18	23	49	29	17	14

Exhibit 7.17   Plato data, clausulae 17-32

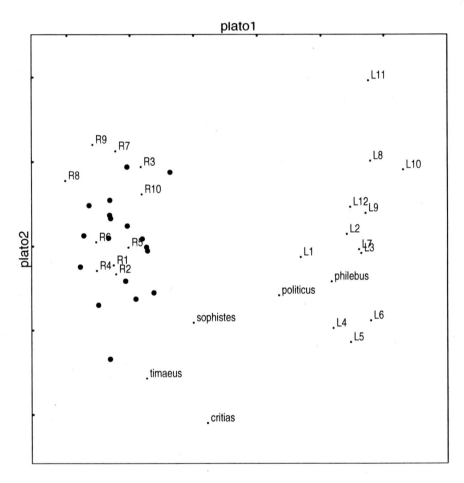

Exhibit 7.18 Multidimensional scaling: Plato data. The fat dots correspond to the 18 dialogues mentioned first place in Exhibits 7.16 and 7.17.

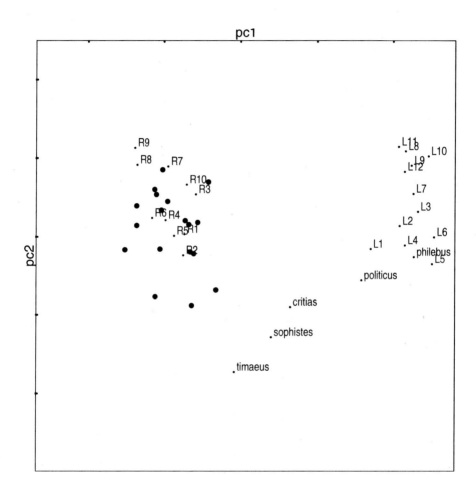

Exhibit 7.19   Correspondence analysis; Plato data. The fat dots correspond to the 18 dialogues mentioned first place in Exhibits 7.16 and 7.17.

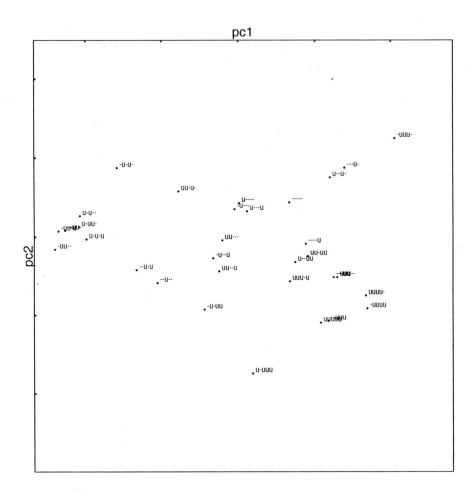

Exhibit 7.20   Correspondence analysis; Plato data. The dots correspond to the clausulae of Exhibits 7.16 and 7.17, mapped into the same space as the works in Exhibit 7.19. In the labels, short syllables are indicated by u, long by –.

Works	Average changes	pc1
R9	2.23	−.28
R8	2.22	−.27
R7	2.20	−.19
lysis	2.17	−.13
R10	2.15	−.14
R6	2.14	−.23
phaedrus	2.14	−.30
euthydemus	2.13	−.27
apology	2.12	−.27
charmides	2.12	−.22
symposium	2.11	−.19
menexenus	2.09	−.21
crito	2.09	−.20
meno	2.08	−.22
R3	2.06	−.12
R4	2.06	−.20
R5	2.05	−.17
protagoras	2.04	−.21
hippias	2.04	−.11
phaedon	2.03	−.14
gorgias	2.02	−.13
R2	2.01	−.15
laches	2.00	−.08
theaetetus	2.00	−.13
parmenides	2.00	−.22
euthyphron	1.98	−.12
R1	1.97	−.15
timaeus	1.96	−.02
cratylus	1.96	−.06
sophistes	1.89	.08
critias	1.87	.13
L12	1.85	.43
L1	1.85	.34
L9	1.84	.45
L10	1.84	.49
L3	1.82	.46
L8	1.80	.43
L11	1.79	.41
L7	1.77	.45
L4	1.77	.43
politicus	1.73	.32
L5	1.72	.50
philebus	1.71	.45
L2	1.69	.42
L6	1.65	.51

Exhibit 7.21   Plato data. Works in reverse order of the average number of changes between short and long syllables in the clausulae; pc1 gives the coordinates of the works in the first principal component direction in Exhibit 7.19. The disputed works are printed in bold face.

CHAPTER 8

MORE CASE STUDIES

The following case studies concern (i) some non-standard examples of dimension reduction through nonlinear local modeling, and (ii) comparison of point configurations. The latter section was added to give at least one example of *quantitative* comparisons, and to round off my discussion of the singular value decomposition with a particularly neat application. I have added a brief outline of numerical optimization.

Assume we have a data set with n items and p variables. Any approach to dimension reduction transforms the space of variables in such a way that the leading variables contain most of the information, while the trailing variables contain mostly noise; subsequently the latter are dropped. A standard approach toward selecting a parsimonious model consists in selecting the most informative variables with the help of a criterion such as Mallows' C_p or Akaike's AIC. This is not necessarily the best strategy since several variables may measure essentially the same quantity, apart from random fluctuations. Then, a weighted sum of those variables will lead to a better model than any single one of them. The question is how to choose that weighted sum. A general approach is through principal components, that is, through

Data Analysis: What Can Be Learned From the Past 50 Years. By Peter J. Huber
Copyright © 2011 John Wiley & Sons, Inc.

the singular value decomposition. The new, lower-dimensional variable space then is a linear transformation of the old one. For a detailed example see Section 7.2.1.

Even more powerful non-standard approaches proceed through local modeling. Here, the idea is to approximate the data pertaining to each individual item by some (nonlinear) model involving as few parameters as feasible. The model parameters then constitute the result of the dimension reduction.

8.1 A NUTSHELL EXAMPLE

This tiny (and somewhat silly) example is given here because it presents in a nutshell the most important stages of a data analysis: Inspection, Error checking, Modification, Comparison, Modeling and Model fitting, Interpretation. Moreover, it prepares for a more complex case of dimension reduction through nonlinear local modeling treated in the following section.

The data were taken from a newspaper report and relate to Ben Johnson's ill-fated 100m world record dash at the World Championship in Rome, August 30, 1987 (the record was later stripped from him because of doping). According to the newspaper (*L'Impartial*, Neuchâtel Sep. 3, 1987), some members of the computer science department at the University of Prague had analyzed the run and obtained the results given in Exhibit 8.1 (raw data taken from the newspaper):

	JOHNSON			LEWIS	
time	mean speed	segment	time	mean speed	
1.86"	19.35 km/h	0–10m	1.94"	18.56 km/h	
1.01"	35.64 km/h	10–20m	1.03"	34.65 km/h	
0.98"	38.71 km/h	20–30m	0.95"	37.89 km/h	
0.86"	41.86 km/h	30–40m	0.85"	42.35 km/h	
0.89"	40.45 km/h	40–50m	0.90"	40.00 km/h	
0.88"	43.37 km/h	50–60m	0.83"	43.37 km/h	
0.83"	43.87 km/h	60–70m	0.83"	43.37 km/h	
0.90"	40.00 km/h	70–80m	0.90"	40.00 km/h	
0.87"	41.38 km/h	80–90m	0.86"	41.86 km/h	
0.85"	42.35 km/h	90–100m	0.84"	42.86 km/h	

Exhibit 8.1 Comparison between Ben Johnson and Carl Lewis, raw data.

The official final times for the 100m dash were: Ben Johnson 9.83" (world record); the second placed Carl Lewis was timed at 9.93". In addition, official reaction times measured at the starting block were given in the newspaper: Johnson 0.129", Lewis 0.196". Reaction times below 0.100" are considered a false start.

Unfortunately, no further details were given in the newspaper. We may assume that the numbers are based on a frame-by-frame analysis of a video tape, presumably with the standard frame rate of 25 fps, or 0.04" between frames.

Data checking. The data contains some obvious misprints, recognizable from the fact that identical mean speeds correspond to different segment times. Moreover, Johnson's segment times add up to 9.93" rather than to 9.83". Fortunately, there is enough redundancy that the errors not only can be recognized as such but also corrected. There are four inconsistencies between segment times and mean speeds; in each case, it suffices to change one digit to achieve agreement.

	JOHNSON			LEWIS	
time	mean speed	segment	time	mean speed
1.86"	19.35 km/h	0–10m	1.94"	18.56 km/h
1.01"	35.64 km/h	10–20m	1.03"	34.65 km/h
0.93"	38.71 km/h	20–30m	0.95"	37.89 km/h
0.86"	41.86 km/h	30–40m	0.85"	42.35 km/h
0.89"	40.45 km/h	40–50m	0.90"	40.00 km/h
0.83"	43.37 km/h	50–60m	0.83"	43.37 km/h
0.83"	43.37 km/h	60–70m	0.83"	43.37 km/h
0.90"	40.00 km/h	70–80m	0.90"	40.00 km/h
0.87"	41.38 km/h	80–90m	0.86"	41.86 km/h
0.85"	42.35 km/h	90–100m	0.84"	42.86 km/h

Exhibit 8.2   Comparison between Ben Johnson and Carl Lewis. Corrected data; the four changed digits are underlined.

Descriptive analysis. It is evident from the tables that the athletes accelerated during the first 30m and then maintained a roughly constant speed till the end of the race, apart from some fluctuations. A comparison between the cumulative times of Johnson and Lewis is interesting:

distance	10	20	30	40	50	60	70	80	90	100
Johnson	1.86	2.87	3.80	4.66	5.55	6.38	7.21	8.11	8.98	9.83
Lewis	1.94	2.97	3.92	4.77	5.67	6.50	7.33	8.23	9.09	9.93
difference	0.08	0.10	0.12	0.11	0.12	0.12	0.12	0.12	0.11	0.10

Thus, in comparison to Johnson, Lewis lost 0.07" (0.196"–0.129" = 0.067") right at the starting block because of a slower reaction time, and another 0.05" in the first 30m of the race. Then he stayed behind by a constant 0.12" (or 1.4m) for the next 50m, and he finally gained back 0.02" in the last 20m.

Modeling and model critique. Presumably, the Prague people had determined the times when the athletes passed 10m marks of the track. Thus, the cumulative rather than the segment times would be affected by independent errors. Because of the relative smallness of those errors (of the order of 1%, or less, of the cumulative times), the error structure of the data is difficult to graph and to ascertain. In order to discern small relative errors, we need to fit a smooth function to the data and to discuss the residuals. It is less important that such a model be physically accurate, than that it is simple and that it does not introduce confusing structure of its own.

Perhaps the simplest smooth model, involving just two free parameters, is to assume that the acceleration $\ddot{x}$ is proportional to the difference between the terminal speed a and the actual speed $\dot{x}$:

$$\ddot{x} = c(a - \dot{x}) \tag{8.1}$$

Integration gives, with the initial conditions $x(0) = \dot{x}(0) = 0$:

$$x(t) = a\left(t + \frac{1}{c}(e^{-ct} - 1)\right). \tag{8.2}$$

A nonlinear least squares fit (either of distance on time, or of time on distance) gave the following RMS residuals (SD) and parameter estimates (with estimated standard errors SE). The reaction times were taken into account. Distances are given in meters, times in seconds:

	distance on time		time on distance	
	Johnson	Lewis	Johnson	Lewis
SD	0.22	0.24	0.019	0.022
a	11.670	11.683	11.668	11.673
SE	0.034	0.038	0.033	0.038
c	0.889	0.851	0.890	0.855
SE	0.013	0.013	0.012	0.013

Exhibit 8.3   Nonlinear least squares fits of the model.

According to the fitted model, the terminal speeds of the two athletes were practically identical ($a = 11.67m/sec$, from fitting time on distance), while the initial acceleration of Johnson was slightly, but significantly, larger ($ac = 10.38m/sec^2$ vs. $9.98m/sec^2$). But a plot of the residuals gave a somewhat surprising picture, see Exhibit 8.4.

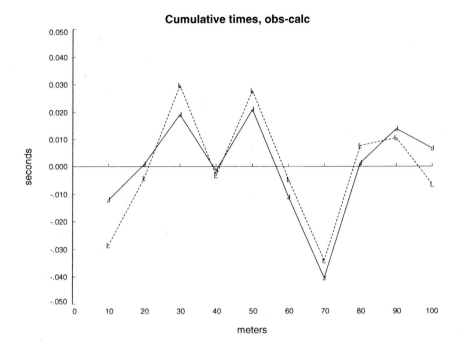

Cumulative times, obs-calc

Exhibit 8.4 Residuals of cumulative times.

The surprise consists in the unexpectedly strong parallelism between the residuals for Johnson and Lewis. We must assume that either the two athletes sped up and slowed down in unison, to a totally implausible degree. Note that in 0.01" the athletes move about 12cm, so the common residuals correspond to values between 25 and 50cm. Or, what is more likely, that there were measurement errors common to both athletes, due to difficulties with the determination of the exact moments when they crossed particular 10m marks. Such errors might be caused by varying view angles of the cameras and the like. Note also that the maximum residual of 0.04" corresponds to the supposed spacing between frames of the videotape. The figure again shows that Johnson had a stronger start and Lewis a stronger finish. Moreover, on closer scrutiny two known inadequacies of the model seem to show up in the figure: our model does not account for the higher initial acceleration, when the athletes are still in contact with the starting blocks, nor for the final spurt. But we lack the data for a confirmatory analysis of these conjectural conclusions.

8.2 SHAPE INVARIANT MODELING

In a longitudinal study conducted at the Children's Hospital at the University of Zurich, extending over 20 years, the development of several hundred children was followed from birth to adulthood. Some of the goals were to produce better tables of normal growth, to help to predict the adult height of a child, and to recognize deviations from the normal growth pattern (to make early hormonal intervention possible).

Human growth is fairly variable from child to child. Exhibit 8.5 shows the salient common patterns with data taken from a small boy. Growth velocity (main part of the figure) is easier to interpret (and to model) than the height itself (top right of the figure – note that the growth curve of the boy moves along the bottom-most percentiles). Unfortunately, the observed velocities are quite noisy, being difference quotients of half-yearly or yearly height measurements (in the figure shown by dots). Before puberty, the growth velocity decreases monotonely. Then there is a pubertal growth spurt, and finally, growth stops near an age of 20 years. Some children lack the growth spurt because of hormonal problems.

Such data present some intrinsic problems. Puberty, and with it the growth spurt, occurs at different ages. If one simply averages growth velocities across the population, the growth spurt is flattened.

Many different approaches are possible and have been tried. A simple one is to apply a spline smoother to individual children. This permits to calculate individual descriptive statistics, in particular three parameters describing the growth spurt: location and height of the peak, and peak-width at half-height, plus two more parameters describing prepubertal growth, such as the velocity at an age of one year, and a measure of the average slope over a few following years.

A more sophisticated approach is to use two model curves, a monotone decreasing one for prepubertal growth, for example:

$$f_1(t) = 1 + \exp(-t),$$

and a bell-shaped one for the growth spurt:

$$f_2(t) = \exp(-t^2).$$

Then one models growth velocity for each individual child i by adding scaled versions of these model curves, with 6 scaling parameters $a_{1i}, b_{1i}, c_{1i}, a_{2i}, b_{2i}, c_{2i}$ estimated from the data for child i:

$$v(t) = a_{1i} f_1((t - b_{1i})/c_{1i}) + a_{2i} f_2((t - b_{2i})/c_{2i}).$$

However, if one now centers the observations at the peak locations and rescales them, then the residuals show systematic deviations from the model curve f_2, and similarly for f_1.

In his Ph.D. thesis, Stützle proposed to improve the model curves to $\hat{f}_1$ and $\hat{f}_2$ by estimating them from the data themselves. He represented these two curves by spline functions with a total of about 30 parameters. There are roughly 400 children with about 36 data points each, and thus one estimates a combined set of roughly $400 \times 6 + 30$ parameters: 6 for each child, and 30 more for the entire population. The computational effort is tolerable because of iterative backfitting: one starts with some initial approximations for $\hat{f}_1$ and $\hat{f}_2$, keeps them fixed and estimates the scaling parameters for each child. Then one fixes these scaling parameters and improves $\hat{f}_1$ and $\hat{f}_2$, and so on.

The estimated model revealed interesting features of the growth process: a mid-spurt at about 7 years, a dip in velocity before puberty and a marked asymmetry of the pubertal peak (Stützle et al. (1980)). Moreover, rather surprisingly, it turned out that the same model curves worked for both boys and girls.

Stützle et al. point out that phenomenologically they are inclined to model the growth process from birth to adulthood with two components, one associated with pubertal and the other with non-pubertal growth. It is mathematically simplest to postulate additivity, but an interaction of the two component is perhaps biologically more realistic: the appearance of pubertal growth would smoothly terminate pre-pubertal growth ('switch-off model'), and they provide evidence that the switch-off model is more consistent with the observations than an additive model. Exhibit 8.5 shows the switch-off model.

Statistically, the fit of the model is excellent, the RMS residuals (about 5 mm) very nearly agree with the independently estimated uncertainties of the measurements.

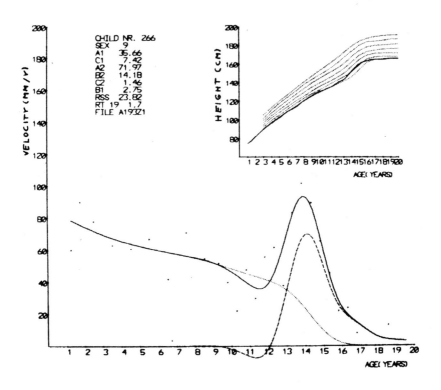

Exhibit 8.5   A typical (modeled) growth curve of a small boy. The figure shows the two components of growth velocity: prepubertal growth with smooth switch-off at puberty, and pubertal growth spurt, and their superposition. The upper right-hand corner shows the cumulative growth of the same small boy. Reproduced with permission, from Stützle et al. (1980: p. 516, Fig. 4).

8.3 COMPARISON OF POINT CONFIGURATIONS

This topic – comparison of spatial point configurations – was included not only because it involves a non-trivial quantitative type of comparison, but also because it occurs in a wide variety of fields, from factor analysis to chemistry, to anthropometry, and to geodesy. An application in marketing research was mentioned in Section 7.4.3. The topic also goes under the name "Procrustes analysis", so named after a figure of Greek mythology who brutally fitted people to a given bed. A distantly related relative is the Thomson problem, with ramifications into pure mathematics, see Section 8.3.2 for a brief discussion.

The problem and its solution go back at least to a paper on factor analysis by Green (1952). A pivotal special case later was used by Golub (1968, p. 45) as an elegant example for the use of the singular value decomposition.

The pivotal problem. Let X and Y be arbitrary $p \times n$ matrices, representing the cartesian coordinates of n points in p-space. The problem is to find an orthogonal $p \times p$ matrix Q such that

$$\|Y - QX\|^2 = \sum_i \sum_j \left(Y_{ji} - \sum_k Q_{jk} X_{ki}\right)^2$$

is minimized.

We assume here that homologous points in the two configurations are numbered identically. If the points are unordered, and if there are many points, it may be quite difficult to put them into one-to-one correspondence. But aligning the configurations along principal component directions, perhaps assisted by projection pursuit, often helps.

The solution is as follows. Determine the singular value decomposition

$$XY^T = USV^T,$$

where U and V are orthogonal $p \times p$ matrices, and S is diagonal with non-negative diagonal elements. Then

$$Q = VU^T,$$

and Q is unique, if and only if X, Y both have full rank p.

This pivotal case extends in a straightforward fashion to similarity transformations, that is, to transformations comprising rotations, inflections, translations and scale changes. Then one has to minimize

$$\|Y - \sigma QX - \eta\|^2,$$

over the matrix Q, the scalar σ and the p-vector η. By centering the configurations at their respective centers of gravity we can get rid of the translation part, because then the optimizing vector η is 0. Furthermore, we note that

$$\|Y - \sigma QX\|^2 = \text{tr}(YY^T) + \sigma^2 \text{tr}(XX^T) - 2\sigma \text{tr}(QXY^T),$$

where tr is the trace function. We conclude first that for any fixed value of σ the optimizing Q is the same, namely the maximizer of $\text{tr}(QXY^T)$. The pivotal recipe thus puts the two configuration into the proper orientation, and it is trivially easy to adjust the scale, namely by

$$\sigma = \mathrm{tr}(QXY^T)/\mathrm{tr}(XX^T).$$

Slightly more difficult problems arise if there are $m \geq 3$ point configurations $Y_1, Y_2, \ldots, Y_m$. Then, as a rule, one will want to determine a reference configuration X from the data, together with the m transformation matrices Q_k, and for that one will use iterative backfitting. Usually, one will then plot the backtransformed configurations $Q_k{}^T Y_k$ on top of the reference configuration X. An example is given in Section 8.3.1.

More complicated families of transformations, such as perspective mappings occurring with photographic data (as in geodesy), seem to require *ad hoc*, brute force approaches through nonlinear least squares. Similarly, if there is need to deal with gross errors or missing values, one will have to use iterative robust methods.

8.3.1 The cyclodecane conformation

Around 1960, the conformations of three different ten-ring molecules had been determined by X-ray analysis in the laboratory of J. D. Dunitz, see Huber-Buser et al. (1961). All three rings consist of 10 carbon atoms, with two nitrogen atoms attached to different carbon atoms. The three molecules differ in crystal symmetry. The surprising facts were first, that the conformation of the ten-ring differed from the "crown" model ("up–down–up–down– ...") which had been used, in the absence of evidence to the contrary, as a working hypothesis in the interpretation of the cyclodecane chemistry. The second surprise was that the three carbon rings, when superimposed on each other, exhibited only very small differences. Exhibit 8.6 shows a superposition of the three molecules, lifted from the original publication. The actual differences are even smaller than what is shown there. A more accurate fit by the methods described above is shown in Exhibit 8.7. The molecules are about 6 Å across, and the maximum discrepancy from an optimally fitted common model is about 0.07 Å. The last figure also illustrates a comment made in Section 2.5.3, namely that visual comparisons are in trouble if the size disparity exceeds 1:30.

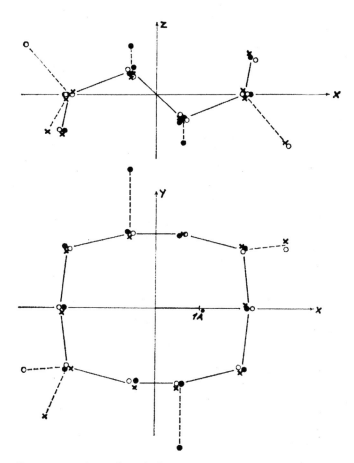

*Representation of cyclodecane ring (with substituents)
as found in* trans-1,6-*diaminocyclodecane dihydro-
chloride, triclinic (●) and monoclinic (○) forms and
in the corresponding* cis-*compound (×). For certain
of the atoms, the distinct points are hardly distinguish-
able from one another.*

Exhibit 8.6   Cyclodecane Ring. From Huber-Buser et al. (1961).

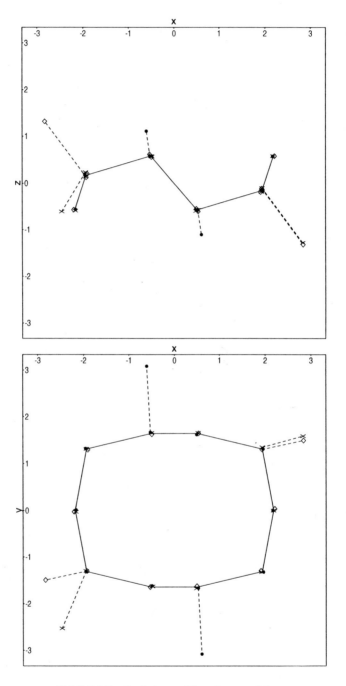

Exhibit 8.7 Cyclodecane Ring. Improved fit.

8.3.2   The Thomson problem

This example was included to give a data analytic example where randomness does not enter at all (compare some comments made at the beginning of Chapter 1).

The Thomson problem is to determine the minimum energy configuration of N electrons on the surface of a sphere that repel each other with a force given by Coulomb's law. It was posed in 1904 by J. J. Thomson (the Nobel prize winning physicist who is credited with the discovery of the electron).

Let $x_1, x_2, \ldots x_N$ be a collection of N distinct points on the unit sphere centered at the origin. The energy of this configuration of points is defined to be

$$\sum_{i<j} \frac{1}{|\mathbf{x}_i - \mathbf{x}_j|^p};$$

$p = 1$ corresponds to the Coulomb potential. The problem is to minimize this energy over all possible collections of N distinct points on the unit sphere. Rigorous solutions are known only for a handful of small values of N.

A neat variant of this problem corresponds to the limiting case $p \to \infty$. It has been posed and investigated by Habicht and van der Waerden (1951) as a problem in pure mathematics: How big must a sphere be that N points with distances ≥ 1 can be fitted onto its surface?

There are local minima, and different values of p may lead to qualitatively different solutions. For example, for $N = 7$ and low values of p, the optimal solution has 5-fold rotational symmetry: it puts one point at each pole and 5 equally spaced points on the equator. But for high values of p the solution has 3-fold rotational symmetry, putting one point on the north pole, three points at a fixed northern latitude, and three other points, rotated by 60°, at a fixed southern latitude. The situation with regard to local minima becomes worse for larger values of N. One may find conjectural configurations by numerical optimization. But comparisons are made difficult by the fact that the points are unordered. If we are to recognize the symmetry properties of an empirically determined configuration, or to compare two configurations that appear to have the same energy, within numerical accuracy, it is necessary to optimize to death. Mere stationarity of the calculated energy does not suffice – it does not determine the configuration to a sufficient accuracy – one must iterate until also the shifts become zero within numerical accuracy. For that, it is necessary to use analytical second order derivatives and to take optimization methods such as Newton's or conjugate gradients. For these methods see the notes in the next section.

8.4 NOTES ON NUMERICAL OPTIMIZATION

Statistical texts and courses usually stay silent on the nitty-gritty of numerical optimization and relegate it to numerical analysis. For a deeper study of numerical optimization see the excellent reference work by Avriel (1976). But the numerical analysis texts notoriously have a learning ramp that is rather steep for someone coming from the outside. The following notes may clarify some of the salient issues. I shall sketch a few general purpose optimization methods of relevance to statistics. The first of them are based on least squares ideas and avoid second order derivatives (whose programming is notoriously inconvenient and error-prone!).

Ordinary linear least squares.
The purpose of ordinary linear least squares is to fit n linear functions $f_i(\theta_1, \ldots, \theta_p) = \sum x_{ij}\theta_j$ of p free parameters to n observations $y_1, \ldots, y_n$. This is done by minimizing

$$\|y - X\theta\|^2 = \sum_{i=1}^{n}\left(y_i - \sum_{j=1}^{p} x_{ij}\theta_j\right)^2 \tag{8.3}$$

A necessary and sufficient condition for $\hat{\theta}$ to minimize this expression is that the residual vector is orthogonal to the column vectors of X, that is $X^T(y - X\hat{\theta}) = 0$, or

$$X^T y - X^T X\hat{\theta} = 0. \tag{8.4}$$

Hence, we have the closed-form solution

$$\hat{\theta} = (X^T X)^{-1} X^T y, \tag{8.5}$$

which, however, is numerically unstable. A stabler solution can be found with the help of the singular value decomposition. Assume that $p \leq n$ and that X has full rank p, and write the SVD as

$$X = U S V^T, \tag{8.6}$$

where U is a $n \times p$ matrix with orthonormal columns, S is $p \times p$ diagonal with positive diagonal elements, and V is orthogonal $p \times p$. Then

$$\hat{\theta} = V S^{-1} U^T y. \tag{8.7}$$

I prefer the SVD to the QR and other decompositions: computationally, it is only slightly more expensive, but it has a better intuitive interpretation, and it gives a neat expression for the hat matrix H that computes the fitted values directly from the observations, $\hat{y} = X\hat{\theta} = Hy$, namely $H = X(X^T X)^{-1} X^T = U U^T$.

The covariance matrix of $\hat{\theta}$ can be estimated as

$$\text{cov}(\hat{\theta}) = \sigma^2 (X^T X)^{-1} = \sigma^2 \, V S^{-2} V^T \tag{8.8}$$

where

$$\sigma^2 = \frac{1}{n-p} \|y - \hat{y}\|^2. \tag{8.9}$$

Nonlinear weighted least squares.
The purpose of nonlinear weighted least squares is to determine $\hat{\theta}$ by minimizing

$$\sum_i w_i \big(y_i - f_i(\theta)\big)^2, \tag{8.10}$$

where the f_i are differentiable functions of $\theta = (\theta_1, \ldots, \theta_p)$. The standard iterative approach avoids the calculation of second order derivatives. Assume that $\theta^{(m)}$ is an approximate solution. Linearize the f_i at $\theta^{(m)}$:

$$f_i(\theta) = f_i(\theta^{(m)}) + \sum_j \frac{\partial}{\partial \theta_j} f_i(\theta^{(m)})(\theta_j - \theta_j^{(m)}). \tag{8.11}$$

Put

$$z_i = \sqrt{w_i}\left(y_i - f_i(\theta^{(m)})\right), \tag{8.12}$$

$$x_{ij} = \sqrt{w_i}\,\frac{\partial}{\partial \theta_j} f_i(\theta^{(m)}), \tag{8.13}$$

$$\eta = \theta - \theta_{(m)}. \tag{8.14}$$

Then solve the ordinary linear least squares problem

$$\|z - X\eta\|^2 = \min! \tag{8.15}$$

and calculate an improved solution

$$\theta^{(m+1)} = \theta^{(m)} + \hat{\eta}. \tag{8.16}$$

Some care is needed because, as with all nonlinear problems, one may get stuck in a local minimum, or the iterative process may fail to converge, especially if it is started from a poor initial value. Also, the analogue of equation (8.9) is valid only if the nonlinearity of the f_i is negligible within the statistical uncertainty range of $\hat{\theta}$.

It is straightforward to robustize linear and nonlinear least squares through the use of *pseudo-observations*, that is by iteratively cutting down large residuals. See Huber and Ronchetti (2009), p. 18-20 for the idea, and p. 170-172 for the necessary bias corrections.

Maximum likelihood.

The general maximum likelihood problem can be reduced to a nonlinear least squares problem. Assume that we want to maximize

$$\sum_{i=1}^{n} \ln f(y_i, \theta) \tag{8.17}$$

where $f(y, \theta)$ is a family of probability densities for a real valued random variable y, depending on a p-dimensional parameter $\theta = (\theta_1, \ldots, \theta_p)$. Ordinarily, we shall assume that the actual distribution of y belongs to the family $f(y, \theta)$. More or less equivalently, we intend to find a θ for which

$$\sum_{i} \psi_j(y_i, \theta) = 0, \quad j = 1, \ldots, p, \tag{8.18}$$

where

$$\psi_j(y, \theta) = \frac{\partial}{\partial \theta_j} \ln f(y, \theta). \tag{8.19}$$

Assume that $\theta^{(m)}$ is an approximate solution and linearize (8.18) there:

$$\sum_{i} \psi_j(y_i, \theta) = \sum_{i} \psi_j(y_i, \theta^{(m)}) + \sum_{k} \left(\sum_{i} \frac{\partial}{\partial \theta_k} \psi_j(y_i, \theta^{(m)}) \right) (\theta_k - \theta_k^{(m)}). \tag{8.20}$$

Now we have essentially three possibilities: we can use Newton's method, we can use conjugate gradients (see below), or we can use a little known variant of nonlinear least squares. The first two converge faster, but are harder to program.

With regard to the third possibility, we note that in a neighborhood of the "true" value θ^0, that is at the value of θ for which the expectation of (8.19) is 0 (whether or not the distribution of y belongs to the family $f(y, \theta)$), the sum

$$\sum_{i} \frac{\partial}{\partial \theta_k} \psi_j(y_i, \theta) \tag{8.21}$$

can be approximated by its expectation, calculated at the true θ^0:

$$n \, \Lambda_{jk} = n \, E\left(\frac{\partial}{\partial \theta_k} \psi_j(y, \theta^0) \right), \tag{8.22}$$

and furthermore, if y is distributed according to $f(y, \theta^0)$, then Λ equals the negative covariance matrix C of the ψ_j:

$$\Lambda_{jk} = -C_{jk} = -E\left(\psi_j(y, \theta^0) \, \psi_k(y, \theta^0) \right). \tag{8.23}$$

In fact, this makes sense even if ψ is not differentiable. See Huber (1967) for technical details and for error bounds. If we replace the last expectation by its sample version, calculated at $\theta^{(m)}$ instead of at the unknown true θ^0, we obtain from (8.20):

$$\sum_i \psi_j(y_i, \theta) = \sum_i \psi_j(y_i, \theta^{(m)}) - \sum_k \left(\sum_i \psi_j(y_i, \theta^{(m)}) \, \psi_k(y_i, \theta^{(m)}) \right) (\theta_k - \theta_k^{(m)}).$$

$$(8.24)$$

The goal is to drive (8.24) to zero. We note that the right hand side is of the same form as (8.4), with $x_{ij} = \psi_j(y_i, \theta^{(m)})$, and the y of (8.4) must be replaced by **1**, where $1_i = 1$ for all i. That is, we simply have to solve the ordinary linear least squares problem

$$\|\mathbf{1} - X\eta\|^2 = \min!$$ $$(8.25)$$

and then put

$$\theta^{(m+1)} = \theta^{(m)} + \hat{\eta}.$$ $$(8.26)$$

This will converge to a solution $\hat{\theta}$ of (8.18). But the convergence behavior can be tricky for small n and for poor starting values, because then the approximation to (8.21) by $n\Lambda$ and of Λ by the sample version of $-C$ can be poor. Some step size control, making sure that (8.17) is increased at each step of the iteration, will be needed to prevent erratic behavior. Asymptotically, the covariance matrix of $\hat{\theta}$ can be estimated by $(X^T X)^{-1}$, provided y is distributed according to $f(y, \theta^0)$; if not, then the asymptotic covariance matrix of $\sqrt{n} \, \hat{\theta}$ is $\Lambda^{-1} C (\Lambda^T)^{-1}$, see Huber (1967).

The general minimization problem.
Assume you want to minimize a twice differentiable function $Q(\mathbf{x})$, where $\mathbf{x} \in \mathbb{R}^p$, and $\mathbf{x}^{(n)}$ is an approximate solution.

The **steepest descent** method improves the solution from $\mathbf{x}^{(n)}$ to $\mathbf{x}^{(n+1)}$ by stepping in the direction of the negative gradient $-\mathbf{v}$ of Q, where $\mathbf{v}$ is defined by $v_j = (\partial/\partial x_j)Q(\mathbf{x})$ at the point $\mathbf{x}^{(n)}$. Consider the function $Q(\mathbf{x}^{(n)} + t\,\mathbf{v})$ of the real variable t and use its first and second derivatives at $t = 0$ to approximate it by a quadratic function. In a neighborhood of the minimum the second derivative is positive, and the quadratic function has a unique minimum. Then let $\mathbf{x}^{(n+1)}$ be the minimizer of the quadratic function. However, after some steep initial steps this converges only slowly.

The **conjugate gradient** method is only slightly more complicated to program but converges considerably faster. Consider the function $Q(\mathbf{x}^{(n)} + s\,\mathbf{u} + t\,\mathbf{v})$ of two real variables s and t, where $\mathbf{u} = \mathbf{x}^{(n)} - \mathbf{x}^{(n-1)}$ is the last step, and $\mathbf{v}$ again is the gradient of Q at $\mathbf{x}^{(n)}$. Approximate this by a quadratic function of s and t, and let $\mathbf{x}^{(n+1)}$ be its minimizer. In the initial stages of the iteration, both steepest descent and conjugate gradients may need some step size control, that is, calculate the value of $Q(\mathbf{x}^{(n+1)})$ and modify the step if necessary, to make sure that Q decreases from step to step. My recommendation is to start the minimization process with steepest descent, and to fall back to a steepest descent step whenever stepsize control kicks in. By the way, it is preferable to use analytic derivatives, whenever possible, rather than numerical ones (in the immediate neighborhood of the minimum the latter become unreliable).

Newton's method uses the full set of first and and second order derivatives of Q at the point $\mathbf{x}^{(n)}$ to approximate Q by a quadratic function of p variables. The next approximation then is defined to be the point $\mathbf{x}^{(n+1)}$ minimizing that quadratic function. If p is large, this becomes computationally expensive. Like steepest descent and conjugate gradients also Newton's method needs some step size control, and annoyingly often there is trouble because the Hessian (the matrix of the second derivatives) fails to be positive definite.

REFERENCES

Avriel, M. (1976). *Nonlinear Programming. Analysis and Methods.* Prentice Hall, NJ. Reprinted by Dover Publishing (2003)

Benzécri, J. P. (1992). *Correspondence Analysis Handbook.* New York, Marcel Dekker.

Besag, J., Green, P., Higdon, D., and Mengersen, K. (1995). Bayesian computations and stochastic systems. *Statistical Science,* **10,** 1-66.

Bickel, P. J., Hammel, E. A., and J. W. O'Connell (1975). Sex Bias in Graduate Admissions: Data from Berkeley. *Science,* **187,** 398-404.

Black, P. (1976). Multidimensional Scaling Applied to Linguistic Relationships. *Cahiers de l'Institut de Linguistique de Louvain, v3 n5-6 Dec 1976,* here cited from "www.eric.ed.gov/ERICWebPortal/recordDetail?accno=ED149581"

Boneva, L. I. (1971). A new approach to a problem of chronological seriation associated with the works of Plato. *Mathematics in the Archaeological and*

Historical Sciences, pp. 173–185, (eds Hodson, F. R., Kendall, D. G., and Tautu, P.). Edinburgh University Press.

Box, G. E. P. (1980). Sampling and Bayes' Inference in Scientific Modelling and Robustness. *J. R. Statist. Soc.* **A 143**, 383-430.

Box, G. E. P. (1990). Comment. *Statistical Science*, **5**, 390-391.

Brainerd, B. (1979). Pronouns and Genre in Shakespeare's Drama. *Computers and the Humanities*, **13**, 3-16.

Breiman, L. (2004). Comment on the NSF Report on the Future of Statistics, *Statistical Science*, **19**, 411.

Breiman, L., Friedman, J. H., Olshen, R. A., and Stone, C. J. (1984). *Classification And Regression Trees.* Chapman and Hall, London.

Brown, E. N. (1988). *Identification and estimation of differential equation models for circadian data.* Ph.D. Thesis, Dept. of Statistics, Harvard University.

Butler, J. N., and Quarrie, D. R. (1996). Data acquisition and analysis in extremely high data rate experiments. *Physics Today,* Oct. 1996, 50-56.

Churchill, W. S. (1951). *The Second World War.* Quoted after the paperback edition, Bantam Books, New York (1962).

Clausewitz, Carl von (1832). *Vom Kriege.* Dümmler Verlag, Bonn. Quoted after the 19th edition (1991).

Clausewitz, Carl von (1984). *On War.* Edited and translated by M. Howard and P. Paret. Princeton University Press, Princeton, NJ.

Coale, A., and Stephan, F. (1962). The Case of the Indians and the Teenage Widows. *J. Amer. Statist. Assoc.* **57**, 338-347.

Cox, D. R., and Brandwood, L. (1959). On a discriminatory problem connected with the works of Plato. *Journal of the Royal Statistical Society.* **B 21**: 195-200.

Cox, D. R., and Hinkley, D. V. (1974). *Theoretical Statistics.* Chapman and Hall, London.

Cowan, C. L., Thompson, W. C., and Ellsworth, E C. (1984). The effects of death qualification on jurors' predisposition to convict and on the quality of deliberation. *Law and Human Behavior*, **8**: 53-80.

Deming, W. E. (1940). Discussion of Professor Hotelling's Paper. *Ann. Math. Statist.,* **11**, 470-471.

De Vaucouleurs, G. (1979). The extragalactic distance scale. *Astrophys. J.,* **227**: 729-755.

Dixon, R. M. W. (1979). Ergativity. *Language*, Vol. **55**, 59-138.

Ehrenberg, A. S. C. (1996). Data Analysis: Strategies and Tactics. *Student* **1**, 271-288. Presses Académiques Neuchâtel.

Fayyad, U. M., Piatetsky-Shapiro, G., Smyth, P., and Uthurusamy, R. (Eds.) (1996). *Advances in Knowledge Discovery and Data Mining*. M.I.T. Press, Cambridge, MA.

Freedman, D.; Pisani, R.; Purves, R., and Adhikari, A. (1991). *Statistics*. 2nd edition. New York: Norton & Company.

French, C. D. (1995). "One Size Fits All" Database Architectures Do Not Work For DSS. *SIGMOD RECORD*, Vol. **24**, June 1995. Proceedings of the 1995 ACM SIGMOD International Conference on Management of Data. ACM Press.

Geman, S. and Geman, D. (1984). Stochastic Relaxation, Gibbs Distributions, and the Bayesian Restoration of Images, *IEEE Transactions on Pattern Analysis and Machine Intelligence*, **6**, 721–741.

Ghurye, S. G., ed. (1975). *Proceedings of the Conference on Directions for Mathematical Statistics*. Special Supplement to *Advances in Applied Probability*.

Golub, G. H. (1968). Least squares, singular values and matrix approximations. *Applicace Matematiky (Applications of Mathematics)*, **13**, 44–51.

Golub, G. H., and Van Loan, C. F. (1983). *Matrix Computations*. The Johns Hopkins University Press. Baltimore, MD.

Green, B. (1952). The orthogonal approximation of an oblique structure in factor analysis. *Psychometrika*, **17**, 429-440.

Habicht, W., und van der Waerden, B. L. (1951). Lagerung von Punkten auf der Kugel. *Mathematische Annalen*, Bd. **123**, 223-234.

Hoare, C. A. R. (1981). The Emperor's Old Clothes. *Comm. of the ACM*, Vol. **24**, 75-83.

Hodson, F. R. (1968). *The La Tène Cemetery at Münsingen-Rain*. Bern: Stämpfli.

Hoeting, J. A., Madigan, D., Raftery, A. E., and Volinsky, C. T. (1999). Bayesian Model Averaging: A Tutorial. *Statistical Science*, **14**, 382-417.

Hotelling, H. (1940). The Teaching of Statistics. *Ann. Math. Statist.*, **11**, 457-470.

Huber, P. J. (1967), The behavior of maximum likelihood estimates under nonstandard conditions, In *Proc. Fifth Berkeley Symposium on Mathematical Statistics and Probability*, Vol. 1, 221–233. University of California Press, Berkeley.

Huber, P. J. (1975a). Applications vs. abstraction: the selling out of mathematical statistics? In: *Proceedings of the Conference on Directions for Mathematical*

Statistics, ed. S. G. Ghurye.   Special Supplement to Advances in Applied Probability.

Huber, P. J. (1975b). Robustness and Designs. In: *A Survey of Statistical Design and Linear Models*, ed. J. N. Srivastava, North-Holland Publishing Company, 287-301.

Huber, P. J. (1981). *Robust Statistics*. Wiley, New York. (See Huber and Ronchetti (2009) for the 2nd edition.)

Huber, P. J. (1985a). Data Analysis: In Search of an Identity. In: *Proc. of the Berkeley Conference in Honor of Jerzy Neyman and Jack Kiefer*, Vol. I. L. M. LeCam and R. A. Olshen (Eds.). Wadsworth, CA.

Huber, P.J. (1985b), Projection Pursuit, *Ann. Statist.*, **13**, 435–475.

Huber, P. J. (1986a). Data Analysis Implications for Command Language Design. In: *Foundation for Human-Computer Communication*. K. Hopper and I. A. Newman (Eds.). Elsevier, North-Holland.

Huber, P. J. (1986b). Environments for supporting statistical strategy. In: *Artificial Intelligence and Statistics*. W. A. Gale (Ed.). Addison-Wesley, Reading, MA.

Huber, P. J. (1994a). Languages for Statistics and Data Analysis. In: *Computational Statistics*, P. Dirschedl and R. Ostermann (Eds.), Physica-Verlag, Heidelberg. (See Huber 2000a for an updated version.)

Huber, P. J. (1994b). Huge Data Sets. In: *Proceedings of the 1994 COMPSTAT Meeting*, R. Dutter and W. Grossmann (Eds.), Physica-Verlag, Heidelberg.

Huber, P. J. (1996a). Massive Data Sets Workshop: The Morning After. In: *Massive Data Sets. Proceedings of a Workshop*. J. Kettenring and D. Pregibon (Eds.). National Academy Press, Washington, D.C.

Huber, P. J. (1996b). *Robust Statistical Procedures* (2nd ed.). SIAM, Philadelphia, PA.

Huber, P. J. (1997a). Speculations on the path of Statistics. In: *The Practice of Data Analysis. Essays in Honor of John W. Tukey*. Brillinger, D. R., Fernholz, L. T., and Morgenthaler, S. (eds.). Princeton University Press, Princeton, NJ.

Huber, P. J. (1997b). Strategy Issues in Data Analysis. In: *Proceedings of the Conference on Statistical Science honoring the bicentennial of Stefano Franscini's birth, Monte Verità, Switzerland*. Malaguerra, C.; Morgenthaler, S., and Ronchetti, E. (eds.). Basel, Birkhäuser Verlag.

Huber, P. J. (1999). Massive Data Sets Workshop: Four Years After. *J. of Computational and Graphical Statistics*, **8**, 635-652.

Huber, P. J. (2000a). Languages for Statistics and Data Analysis. *J. of Computational and Graphical Statistics*, **9**, 600-620. (Updated version of Huber 1994a).

Huber, P. J. (2000b). Babylonian short-time measurements. *Centaurus*. **42**, 223-234.

Huber, P. J. (2002). Approximate Models. In: *Goodness-of-Fit Tests and Model Validity*. C. Huber-Carol, N. Balakrishnan, M. S. Nikulin and M. Mesbah (eds.). Basel, Birkhäuser Verlag.

Huber, P. J. (2006). Modeling the length of day and extrapolating the rotation of the Earth. *Journal of Geodesy*, **80**, 283-303.

Huber, P. J. (2011). Dating of Akkad, Ur III and Babylon I. *Proceedings of the 54th Rencontre Assyriologique Internationale 2008*. (In press.)

Huber, P. J., and De Meis, S. (2004). *Babylonian Eclipse Observations from 750 BC to 1 BC*. IsIAO - Mimesis, Milano.

Huber, P. J., and Huber-Buser, E. H. (1988). ISP: Why a Command Language? In: *Fortschritte der Statistik-Software I*, F. Faulbaum und H.-M. Uehlinger (Hrsg.). Gustav Fischer, Stuttgart.

Huber, P. .J., and Ronchetti, E. M. (2009). *Robust Statistics*, 2nd. edition, Wiley, New York.

Huber, P. J., Sachs, A., Stol, M., Whiting, R. M., Leichty, E., Walker, C. B. F. and vanDriel, G. (1982). *Astronomical Dating of Babylon I and Ur III*. Occasional Papers on the Near East, Vol. 1, Issue 4. Malibu: Undena Publications.

Huber, Th. M., and Nagel, M. (1996). Data Based Prototyping. In: *Robust Statistics, Data Analysis, and Computer Intensive Methods*, H. Rieder (Ed.), Springer, New York.

Huber-Buser, E., Dunitz, J. D., and Venkatesan, K. (1961). Conformation of the Cyclodecane Ring. *Proc. Chem. Soc.*, 463.

Iverson, K. E. (1962). *A Programming Language*. Wiley, New York.

Jalics, P. J., and Heines, T. S. (1983). Transporting a portable operating system: UNIX to an IBM minicomputer. *Comm. of the ACM*, Vol. **26**, 1066-1072.

Johnson, J. A., Nardi, B. A., Zarmer, C. L., and Miller, J. R. (1993). ACE: Building interactive graphical applications. *Comm. of the ACM*, Vol. **36**, No. 4, 41-55.

Jordi, C., Morrison, L. V., Rosen, R. D., Salstein, D. A., Rosselló, G. (1994). Fluctuations in the Earth's rotation since 1830 from high-resolution astronomical data. *Geophys. J. Int.* **117**:811-818.

Kaluscha, W. (1904). Zur Chronologie der platonischen Dialoge. *Wiener Studien*, pp. 25-27.

Kant, I. (1997). *Prolegomena to Any Future Metaphysics.* Translated and edited by Gary Hatfield. Cambridge University Press.

Kendall, D. G. (1970). A mathematical approach to seriation. *Phil. Trans. R. Soc. Lond.* **A 269**:125-135.

Kendall, D. G. (1971). Seriation from abundance matrices. *Mathematics in the Archaeological and Historical Sciences,* pp. 115–287, (eds Hodson, F. R., Kendall, D. G., and Tautu, P.). Edinburgh University Press.

Kettenring, J. and Pregibon, D. (Eds.) (1996). *Massive Data Sets. Proceedings of a Workshop.* National Academy Press, Washington, D.C.

Knuth, D. (1968-). *The Art of Computer Programming.* Addison-Wesley, Reading MA.

Knuth, D. (1986). *Computers and Typesetting.* Addison-Wesley, Reading MA.

Kruskal, J. B. (1964). Nonmetric multidimensional scaling. *Psychometrika* **29**, 1-27, 115-129.

Lander, E. S. (1995). Mapping heredity: Using probabilistic models and algorithms to map genes and genomes, *Notices of the AMS,* July 1995, 747-753. Adapted from: *Calculating the Secrets of Life.* National Academy Press, Washinton, D.C. 1995.

Larsen, M. T. (1976). *The Old Assyrian City-State and its Colonies.* Mesopotamia 4, Akademisk Forlag, Copenhagen.

Lehmann, E. L. (1986). *Testing Statistical Hypotheses.* 2nd edition. Wiley, New York.

Lehmann, W. P. (1978). *Syntactic Typology. Studies in the Phenomenology of Language.* Univ. of Texas Press, Austin.

Litecky, C. R., and Davis, G. B. (1976). A study of errors, error-proneness, and error diagnosis in Cobol. *Comm. of the ACM,* Vol. **19**, 33-37.

Little, R. J. A., and Rubin, D. B. (1987). Statistical Analysis with Missing Data. Wiley, New York.

Mallows, C. (2006). Tukey's Paper After 40 Years. *Technometrics,* **48**, 319–336.

Mao Tse-tung (1937). On Contradiction. Quoted from *Mao Tse-tung: An Anthology of his Writings,* edited by Anne Freemantle, The New American Library (1962), 240.

McCullagh, P., and Nelder, J. A. (1983). *Generalized Linear Models.* Chapman and Hall, London.

Miller, B. P., Fredriksen L. and So, B. (1990). An empirical study of the reliability of UNIX utilities. *Comm. of the ACM*, Vol. 33, No. 12, 32-42.

Muller, P. M., and Sjogren, W. L. (1968). Mascons: Lunar Mass Concentrations. *Science*, **161**, 680-684.

Munk, W. H., and Snodgrass, F.E. (1957). Measurements of southern swell at Guadeloupe Island. Deep-Sea Research, **4**, 272-286.

Naur, P. (1960). Report on the algorithmic language ALGOL 60. *Comm. of the ACM*, Vol.**3**, 299-314.

Naur, P. (1975). Programming languages, natural languages, and mathematics. *Comm. of the ACM*, Vol. **18**, 676-682.

Parrish, W. (1960). Results of the I.U.Cr. precision lattice parameter project. *Acta Cryst.* **13**, 838-850.

Pearson, K. (1900). On the criterion that a given system of deviations from the probable in the case of a correlated system of variables is such that it can be reasonably supposed to have arisen from random sampling. *Philosophical Magazine*, **50**, 157-172.

Playfair, W. (1821). *A letter on our agricultural distresses, their causes and remedies.* London.

Postscript Language. *Tutorial and Cookbook.* (1985). Adobe Systems, Inc.

Rawal, K. (1992). A macro facility for X. *The X Resource 1, Winter 1992, 6th Annual X Technical Conference.* p. 133-142.

Rutishauser, H. (1952). Automatische Rechenplanfertigung bei programmgesteuerten Rechenmaschinen. *Mitt. Inst. Angewandte Math. ETH Zürich.* Nr. 3. Birkhäuser, Basel/Stuttgart.

Schmid, P. (1967). On "Grossversuch III", a randomized hail suppression experiment in Switzerland. In: *Proc. of the Fifth Berkeley Symposium,* Vol. **V**. Univ. of Calif. Press.

Sedransk. N., Young., L. J., Kelner, K. L., Moffitt, R. A., Thakar, A., Raddick, J., Ungvarsky, E. J., Carlson, R. W., Apweiler, R., Cox, L. H., Nolan, D., Soper, K., and Spiegelman, C. (2010). Make Research Data Public? – Not Always so Simple: A Dialogue for Statisticians and Science Editors. *Statistical Science.* **25**, 41-50.

Shannon, C. E. (1949). *The Mathematical Theory of Communication.* U. of Illinois Press.

Simpson, E. H. (1951). The interpretation of interaction in contingency tables. *Journal of the Royal Statistical Society.* **B 13**: 238241.

Slutsky, E. (1927) The summation of random causes as the source of cyclic processes (in Russian), *Problems of Economic Conditions*, **3**:1. English trans. in *Econometrica*, **5**:105 (1937).

Stephenson, F. R. and Morrison, L. V. (1995). Long-term fluctuations in the Earth's rotation: 700 BC to AD 1990. *Phil. Trans. R. Soc. Lond.* **A 351**:165-202

Stuetzle, W. (1987). Plot Windows. *J. Amer. Statist. Assoc.*, Vol. **82**, 466-475.

Stuetzle, W., Gasser, T., Molinari, L., Largo, R. H., Prader, A., and Huber, P. J. (1980). Shape-invariant modelling of human growth. *Annals of Human Biology*, Vol. **7**, No. 6, 507–528.

Sun Tzu (1963). *The Art of War.* Translated and with an introduction by Samuel B. Griffith. Oxford Univ. Press.

Taubes, G. (1996). Redefining the Supercomputer. *Science*, **273**, 1655-1657.

Thompson, W.C., Cowan, C. L., Ellsworth, P.C., and Harrington, J.C. (1984). Death penalty attitudes and conviction proneness. The translation of attitudes into verdicts. *Law and Human Behavior*, **8**, 95-113.

Thomson, J. J. (1904). On the Structure of the Atom: an Investigation of the Stability and Periods of Oscillation of a number of Corpuscles arranged at equal intervals around the Circumference of a Circle; with Application of the Results to the Theory of Atomic Structure. *Philosophical Magazine.* Series 6, Volume 7, Number 39, 237-265.

Tufte, E. R. (1983). *The Visual Display of Quantitative Information.* Graphics Press, Cheshire CT.

Tukey, J. W. (1962). The Future of Data Analysis. *Ann. Math. Statist.*, **33**, 1-67.

Tukey, J. W. (1965). Use of numerical spectrum analysis in geophysics. *Bulletin of the International Statistical Institute. Proceedings of the 35th Session, Beograd 1965)*

Tukey, J. W. (1968). Is Statistics a Computing Science? In: *The Future of Statistics*, ed. D. G. Watts. Academic Press, New York and London.

Tukey, J.W. (1970), *Exploratory Data Analysis*, Mimeographed Preliminary Edition.

Tukey, J.W. (1977), *Exploratory Data Analysis*, Addison-Wesley, Reading, MA.

UNIX Time Sharing System (1978). *Bell Systems Technical Journal,* Vol. **57** (1978), No. 6, Part 2.

Vach, W. (1987). The design of a 'Lab Assistant' for data analysis. *Tech. Report.*

Veenhof, K. R. (2003). *The Old Assyrian List of Year Eponyms from Karum Kanish and its Chronological Implications.* Türk Tarih Kurumu, Ankara.

Watts, D. G., ed. (1968). *The Future of Statistics. Proceedings of a Conference on the Future of Statistics Held at the University of Wisconsin, Madison.* Academic Press, New York and London.

Wegman, E. (1995). Huge data sets and the frontiers of computational feasibility. *J. of Computational and Graphical Statistics,* **4**, 281-295.

Wiener, N. (1963). *Cybernetics,* 2nd. edition. M.I.T. Press, Cambridge, MA.

Wirth, N. (1971). The programming language Pascal. *Acta Informatica,* Vol.**1**, 35-63.

Wirth, N. (1985). From programming language design to computer construction. *Comm. of the ACM,* Vol. **28**, 160-164.

X Window (1993). *The Definitive Guide to the X Window System* (various authors). O'Reilly & Associates, Inc., Sebastopol, CA.

INDEX

Data Analysis: What Can Be Learned From the Past 50 Years. By Peter J. Huber
Copyright © 2011 John Wiley & Sons, Inc.

WILEY SERIES IN PROBABILITY AND STATISTICS
ESTABLISHED BY WALTER A. SHEWHART AND SAMUEL S. WILKS

Editors: *David J. Balding, Noel A. C. Cressie, Garrett M. Fitzmaurice, Iain M. Johnstone, Geert Molenberghs, David W. Scott, Adrian F. M. Smith, Ruey S. Tsay, Sanford Weisberg*
Editors Emeriti: *Vic Barnett, J. Stuart Hunter, Joseph B. Kadane, Jozef L. Teugels*

The *Wiley Series in Probability and Statistics* is well established and authoritative. It covers many topics of current research interest in both pure and applied statistics and probability theory. Written by leading statisticians and institutions, the titles span both state-of-the-art developments in the field and classical methods.

Reflecting the wide range of current research in statistics, the series encompasses applied, methodological and theoretical statistics, ranging from applications and new techniques made possible by advances in computerized practice to rigorous treatment of theoretical approaches.

This series provides essential and invaluable reading for all statisticians, whether in academia, industry, government, or research.

† ABRAHAM and LEDOLTER · Statistical Methods for Forecasting
AGRESTI · Analysis of Ordinal Categorical Data, *Second Edition*
AGRESTI · An Introduction to Categorical Data Analysis, *Second Edition*
AGRESTI · Categorical Data Analysis, *Second Edition*
ALTMAN, GILL, and McDONALD · Numerical Issues in Statistical Computing for the Social Scientist
AMARATUNGA and CABRERA · Exploration and Analysis of DNA Microarray and Protein Array Data
ANDĚL · Mathematics of Chance
ANDERSON · An Introduction to Multivariate Statistical Analysis, *Third Edition*
* ANDERSON · The Statistical Analysis of Time Series
ANDERSON, AUQUIER, HAUCK, OAKES, VANDAELE, and WEISBERG · Statistical Methods for Comparative Studies
ANDERSON and LOYNES · The Teaching of Practical Statistics
ARMITAGE and DAVID (editors) · Advances in Biometry
ARNOLD, BALAKRISHNAN, and NAGARAJA · Records
* ARTHANARI and DODGE · Mathematical Programming in Statistics
* BAILEY · The Elements of Stochastic Processes with Applications to the Natural Sciences
BALAKRISHNAN and KOUTRAS · Runs and Scans with Applications
BALAKRISHNAN and NG · Precedence-Type Tests and Applications
BARNETT · Comparative Statistical Inference, *Third Edition*
BARNETT · Environmental Statistics
BARNETT and LEWIS · Outliers in Statistical Data, *Third Edition*
BARTOSZYNSKI and NIEWIADOMSKA-BUGAJ · Probability and Statistical Inference
BASILEVSKY · Statistical Factor Analysis and Related Methods: Theory and Applications
BASU and RIGDON · Statistical Methods for the Reliability of Repairable Systems
BATES and WATTS · Nonlinear Regression Analysis and Its Applications

*Now available in a lower priced paperback edition in the Wiley Classics Library.
†Now available in a lower priced paperback edition in the Wiley–Interscience Paperback Series.

*Now available in a lower priced paperback edition in the Wiley Classics Library.

†Now available in a lower priced paperback edition in the Wiley–Interscience Paperback Series.

CHOW and LIU · Design and Analysis of Clinical Trials: Concepts and Methodologies, *Second Edition*

CLARKE · Linear Models: The Theory and Application of Analysis of Variance

CLARKE and DISNEY · Probability and Random Processes: A First Course with Applications, *Second Edition*

* COCHRAN and COX · Experimental Designs, *Second Edition*

COLLINS and LANZA · Latent Class and Latent Transition Analysis: With Applications in the Social, Behavioral, and Health Sciences

CONGDON · Applied Bayesian Modelling

CONGDON · Bayesian Models for Categorical Data

CONGDON · Bayesian Statistical Modelling

CONOVER · Practical Nonparametric Statistics, *Third Edition*

COOK · Regression Graphics

COOK and WEISBERG · Applied Regression Including Computing and Graphics

COOK and WEISBERG · An Introduction to Regression Graphics

CORNELL · Experiments with Mixtures, Designs, Models, and the Analysis of Mixture Data, *Third Edition*

COVER and THOMAS · Elements of Information Theory

COX · A Handbook of Introductory Statistical Methods

* COX · Planning of Experiments

CRESSIE · Statistics for Spatial Data, *Revised Edition*

CRESSIE and WIKLE · Statistics for Spatio-Temporal Data

CSÖRGŐ and HORVÁTH · Limit Theorems in Change Point Analysis

DANIEL · Applications of Statistics to Industrial Experimentation

DANIEL · Biostatistics: A Foundation for Analysis in the Health Sciences, *Eighth Edition*

* DANIEL · Fitting Equations to Data: Computer Analysis of Multifactor Data, *Second Edition*

DASU and JOHNSON · Exploratory Data Mining and Data Cleaning

DAVID and NAGARAJA · Order Statistics, *Third Edition*

* DEGROOT, FIENBERG, and KADANE · Statistics and the Law

DEL CASTILLO · Statistical Process Adjustment for Quality Control

DeMARIS · Regression with Social Data: Modeling Continuous and Limited Response Variables

DEMIDENKO · Mixed Models: Theory and Applications

DENISON, HOLMES, MALLICK and SMITH · Bayesian Methods for Nonlinear Classification and Regression

DETTE and STUDDEN · The Theory of Canonical Moments with Applications in Statistics, Probability, and Analysis

DEY and MUKERJEE · Fractional Factorial Plans

DILLON and GOLDSTEIN · Multivariate Analysis: Methods and Applications

DODGE · Alternative Methods of Regression

* DODGE and ROMIG · Sampling Inspection Tables, *Second Edition*

* DOOB · Stochastic Processes

DOWDY, WEARDEN, and CHILKO · Statistics for Research, *Third Edition*

DRAPER and SMITH · Applied Regression Analysis, *Third Edition*

DRYDEN and MARDIA · Statistical Shape Analysis

DUDEWICZ and MISHRA · Modern Mathematical Statistics

DUNN and CLARK · Basic Statistics: A Primer for the Biomedical Sciences, *Third Edition*

DUPUIS and ELLIS · A Weak Convergence Approach to the Theory of Large Deviations

EDLER and KITSOS · Recent Advances in Quantitative Methods in Cancer and Human Health Risk Assessment

* ELANDT-JOHNSON and JOHNSON · Survival Models and Data Analysis

*Now available in a lower priced paperback edition in the Wiley Classics Library.
†Now available in a lower priced paperback edition in the Wiley–Interscience Paperback Series.

ENDERS · Applied Econometric Time Series

† ETHIER and KURTZ · Markov Processes: Characterization and Convergence

EVANS, HASTINGS, and PEACOCK · Statistical Distributions, *Third Edition*

FELLER · An Introduction to Probability Theory and Its Applications, Volume I, *Third Edition,* Revised; Volume II, *Second Edition*

FISHER and VAN BELLE · Biostatistics: A Methodology for the Health Sciences

FITZMAURICE, LAIRD, and WARE · Applied Longitudinal Analysis

* FLEISS · The Design and Analysis of Clinical Experiments

FLEISS · Statistical Methods for Rates and Proportions, *Third Edition*

† FLEMING and HARRINGTON · Counting Processes and Survival Analysis

FUJIKOSHI, ULYANOV, and SHIMIZU · Multivariate Statistics: High-Dimensional and Large-Sample Approximations

FULLER · Introduction to Statistical Time Series, *Second Edition*

† FULLER · Measurement Error Models

GALLANT · Nonlinear Statistical Models

GEISSER · Modes of Parametric Statistical Inference

GELMAN and MENG · Applied Bayesian Modeling and Causal Inference from Incomplete-Data Perspectives

GEWEKE · Contemporary Bayesian Econometrics and Statistics

GHOSH, MUKHOPADHYAY, and SEN · Sequential Estimation

GIESBRECHT and GUMPERTZ · Planning, Construction, and Statistical Analysis of Comparative Experiments

GIFI · Nonlinear Multivariate Analysis

GIVENS and HOETING · Computational Statistics

GLASSERMAN and YAO · Monotone Structure in Discrete-Event Systems

GNANADESIKAN · Methods for Statistical Data Analysis of Multivariate Observations, *Second Edition*

GOLDSTEIN and LEWIS · Assessment: Problems, Development, and Statistical Issues

GREENWOOD and NIKULIN · A Guide to Chi-Squared Testing

GROSS, SHORTLE, THOMPSON, and HARRIS · Fundamentals of Queueing Theory, *Fourth Edition*

GROSS, SHORTLE, THOMPSON, and HARRIS · Solutions Manual to Accompany Fundamentals of Queueing Theory, *Fourth Edition*

* HAHN and SHAPIRO · Statistical Models in Engineering

HAHN and MEEKER · Statistical Intervals: A Guide for Practitioners

HALD · A History of Probability and Statistics and their Applications Before 1750

HALD · A History of Mathematical Statistics from 1750 to 1930

† HAMPEL · Robust Statistics: The Approach Based on Influence Functions

HANNAN and DEISTLER · The Statistical Theory of Linear Systems

HARTUNG, KNAPP, and SINHA · Statistical Meta-Analysis with Applications

HEIBERGER · Computation for the Analysis of Designed Experiments

HEDAYAT and SINHA · Design and Inference in Finite Population Sampling

HEDEKER and GIBBONS · Longitudinal Data Analysis

HELLER · MACSYMA for Statisticians

HINKELMANN and KEMPTHORNE · Design and Analysis of Experiments, Volume 1: Introduction to Experimental Design, *Second Edition*

HINKELMANN and KEMPTHORNE · Design and Analysis of Experiments, Volume 2: Advanced Experimental Design

HOAGLIN, MOSTELLER, and TUKEY · Fundamentals of Exploratory Analysis of Variance

* HOAGLIN, MOSTELLER, and TUKEY · Exploring Data Tables, Trends and Shapes

* HOAGLIN, MOSTELLER, and TUKEY · Understanding Robust and Exploratory Data Analysis

*Now available in a lower priced paperback edition in the Wiley Classics Library.

†Now available in a lower priced paperback edition in the Wiley–Interscience Paperback Series.

*Now available in a lower priced paperback edition in the Wiley Classics Library.

†Now available in a lower priced paperback edition in the Wiley–Interscience Paperback Series.

*Now available in a lower priced paperback edition in the Wiley Classics Library.
†Now available in a lower priced paperback edition in the Wiley–Interscience Paperback Series.

RENCHER · Methods of Multivariate Analysis, *Second Edition*
RENCHER · Multivariate Statistical Inference with Applications
* RIPLEY · Spatial Statistics
* RIPLEY · Stochastic Simulation
ROBINSON · Practical Strategies for Experimenting
ROHATGI and SALEH · An Introduction to Probability and Statistics, *Second Edition*
ROLSKI, SCHMIDLI, SCHMIDT, and TEUGELS · Stochastic Processes for Insurance and Finance
ROSENBERGER and LACHIN · Randomization in Clinical Trials: Theory and Practice
ROSS · Introduction to Probability and Statistics for Engineers and Scientists
ROSSI, ALLENBY, and McCULLOCH · Bayesian Statistics and Marketing
† ROUSSEEUW and LEROY · Robust Regression and Outlier Detection
* RUBIN · Multiple Imputation for Nonresponse in Surveys
RUBINSTEIN and KROESE · Simulation and the Monte Carlo Method, *Second Edition*
RUBINSTEIN and MELAMED · Modern Simulation and Modeling
RYAN · Modern Engineering Statistics
RYAN · Modern Experimental Design
RYAN · Modern Regression Methods, *Second Edition*
RYAN · Statistical Methods for Quality Improvement, *Second Edition*
SALEH · Theory of Preliminary Test and Stein-Type Estimation with Applications
* SCHEFFE · The Analysis of Variance
SCHIMEK · Smoothing and Regression: Approaches, Computation, and Application
SCHOTT · Matrix Analysis for Statistics, *Second Edition*
SCHOUTENS · Levy Processes in Finance: Pricing Financial Derivatives
SCHUSS · Theory and Applications of Stochastic Differential Equations
SCOTT · Multivariate Density Estimation: Theory, Practice, and Visualization
† SEARLE · Linear Models for Unbalanced Data
† SEARLE · Matrix Algebra Useful for Statistics
† SEARLE, CASELLA, and McCULLOCH · Variance Components
SEARLE and WILLETT · Matrix Algebra for Applied Economics
SEBER · A Matrix Handbook For Statisticians
† SEBER · Multivariate Observations
SEBER and LEE · Linear Regression Analysis, *Second Edition*
† SEBER and WILD · Nonlinear Regression
SENNOTT · Stochastic Dynamic Programming and the Control of Queueing Systems
* SERFLING · Approximation Theorems of Mathematical Statistics
SHAFER and VOVK · Probability and Finance: It's Only a Game!
SILVAPULLE and SEN · Constrained Statistical Inference: Inequality, Order, and Shape Restrictions
SMALL and McLEISH · Hilbert Space Methods in Probability and Statistical Inference
SRIVASTAVA · Methods of Multivariate Statistics
STAPLETON · Linear Statistical Models, *Second Edition*
STAPLETON · Models for Probability and Statistical Inference: Theory and Applications
STAUDTE and SHEATHER · Robust Estimation and Testing
STOYAN, KENDALL, and MECKE · Stochastic Geometry and Its Applications, *Second Edition*
STOYAN and STOYAN · Fractals, Random Shapes and Point Fields: Methods of Geometrical Statistics
STREET and BURGESS · The Construction of Optimal Stated Choice Experiments: Theory and Methods
STYAN · The Collected Papers of T. W. Anderson: 1943–1985
SUTTON, ABRAMS, JONES, SHELDON, and SONG · Methods for Meta-Analysis in Medical Research
TAKEZAWA · Introduction to Nonparametric Regression

*Now available in a lower priced paperback edition in the Wiley Classics Library.
†Now available in a lower priced paperback edition in the Wiley–Interscience Paperback Series.

TAMHANE · Statistical Analysis of Designed Experiments: Theory and Applications

TANAKA · Time Series Analysis: Nonstationary and Noninvertible Distribution Theory

THOMPSON · Empirical Model Building

THOMPSON · Sampling, *Second Edition*

THOMPSON · Simulation: A Modeler's Approach

THOMPSON and SEBER · Adaptive Sampling

THOMPSON, WILLIAMS, and FINDLAY · Models for Investors in Real World Markets

TIAO, BISGAARD, HILL, PEÑA, and STIGLER (editors) · Box on Quality and Discovery: with Design, Control, and Robustness

TIERNEY · LISP-STAT: An Object-Oriented Environment for Statistical Computing and Dynamic Graphics

TSAY · Analysis of Financial Time Series, *Third Edition*

UPTON and FINGLETON · Spatial Data Analysis by Example, Volume II: Categorical and Directional Data

† VAN BELLE · Statistical Rules of Thumb, *Second Edition*

VAN BELLE, FISHER, HEAGERTY, and LUMLEY · Biostatistics: A Methodology for the Health Sciences, *Second Edition*

VESTRUP · The Theory of Measures and Integration

VIDAKOVIC · Statistical Modeling by Wavelets

VINOD and REAGLE · Preparing for the Worst: Incorporating Downside Risk in Stock Market Investments

WALLER and GOTWAY · Applied Spatial Statistics for Public Health Data

WEERAHANDI · Generalized Inference in Repeated Measures: Exact Methods in MANOVA and Mixed Models

WEISBERG · Applied Linear Regression, *Third Edition*

WEISBERG · Bias and Causation: Models and Judgment for Valid Comparisons

WELSH · Aspects of Statistical Inference

WESTFALL and YOUNG · Resampling-Based Multiple Testing: Examples and Methods for *p*-Value Adjustment

WHITTAKER · Graphical Models in Applied Multivariate Statistics

WINKER · Optimization Heuristics in Economics: Applications of Threshold Accepting

WONNACOTT and WONNACOTT · Econometrics, *Second Edition*

WOODING · Planning Pharmaceutical Clinical Trials: Basic Statistical Principles

WOODWORTH · Biostatistics: A Bayesian Introduction

WOOLSON and CLARKE · Statistical Methods for the Analysis of Biomedical Data, *Second Edition*

WU and HAMADA · Experiments: Planning, Analysis, and Parameter Design Optimization, *Second Edition*

WU and ZHANG · Nonparametric Regression Methods for Longitudinal Data Analysis

YANG · The Construction Theory of Denumerable Markov Processes

YOUNG, VALERO-MORA, and FRIENDLY · Visual Statistics: Seeing Data with Dynamic Interactive Graphics

ZACKS · Stage-Wise Adaptive Designs

ZELTERMAN · Discrete Distributions—Applications in the Health Sciences

* ZELLNER · An Introduction to Bayesian Inference in Econometrics

ZHOU, OBUCHOWSKI, and McCLISH · Statistical Methods in Diagnostic Medicine, *Second Edition*

*Now available in a lower priced paperback edition in the Wiley Classics Library.

†Now available in a lower priced paperback edition in the Wiley–Interscience Paperback Series.